Bridge Bearings and Expansion Joints

Other Titles From E & FN Spon

Alternative Materials for the Reinforcement and Prestressing of Concrete
Edited by J.L. Clarke

Architecture and Construction in Steel
Edited by A. Blanc, M. McEvoy and R. Plank

Bridge Deck Behaviour
E.C. Hambly

Concrete Bridge Designer's Manual
E. Pennells

Concrete Bridge Engineering: Performance and Advances
R.J. Cope

Concrete Structures: Stresses and Deformations
A. Ghali and R. Favre

Construction Methods and Planning
J.R. Illingworth

Cyclic Loading of Soils
M.P. O'Reilly and S.F. Brown

Design of Prestressed Concrete
R.I. Gilbert and N.C. Mickleborough

Earth Pressure and Earth-Retaining Structures
C.R.I. Clayton, J. Milititsky and R.I. Woods

Engineering Treatment of Soils
F.G. Bell

Flexural-Torsional Buckling of Structures
N.S. Trahair

Foundations on Rock
D.C. Wyllie

High Performance Concrete: From Material to Structure
Edited by Y. Malier

Integral Bridges
Edited by B. Pritchard

Pile Design and Construction Practice
M.J. Tomlinson

Piling Engineering
W.G.K. Fleming, A.J. Weltman, M.F. Randolf and W.K. Elson

Reinforced Concrete Designer's Handbook
C.E. Reynolds and J.C. Steedman

Structural Dynamics for the Practising Engineer
H.M. Irvine

The Design Life of Structures
Edited by G. Somerville

The Maintenance of Brick and Stone Masonry Structures
Edited by A.M. Sowden

For information about these and other titles, please contact:
The Promotion Department, E & FN Spon,
2–6 Boundary Row, London SE1 8HN, Tel: 071 522 9966

Bridge Bearings and Expansion Joints

Second edition

David J. Lee
G. Maunsell & Partners
Kent
UK

E & FN SPON
An Imprint of Chapman & Hall

London · Glasgow · New York · Tokyo · Melbourne · Madras

Published by E & FN Spon, an imprint of Chapman & Hall, 2–6 Boundary Row, London SE1 8HN, UK

Chapman & Hall, 2–6 Boundary Row, London SE1 8HN, UK

Chapman & Hall GmbH Pappelallee 3, 69469 Weinheim, Germany

Blackie Academic & Professional, Wester Cleddens Road, Bishopbriggs, Glasgow G64 2NZ, UK

Chapman & Hall Inc., One Penn Plaza, New York NY10119, USA

Chapman & Hall Japan, Thomson Publishing Japan, Hirakawacho Nemoto Building, 6F, 1-7-11 Hirakawa-cho, Chiyoda-ku, Tokyo 102, Japan

Chapman & Hall Australia, Thomas Nelson Australia, 102 Dodds Street, South Melbourne, Victoria 3205, Australia

Chapman & Hall India, R. Seshadri, 32 Second Main Road, CIT East, Madras 600 035, India

First published in 1971 by the Cement and Concrete Association
Second edition 1994.

© 1971 Cement and Concrete Association
© 1994 David J. Lee
Designed by Geoffrey Wadsley
Typeset in 10/12pt Times by Expo Holdings, Malaysia
Printed in Great Britain by the Alden Press, Osney Mead, Oxford

ISBN 0 419 14570 2

A catalogue record for this book is available from the British Library

Library of Congress Cataloging-in-Publication
Lee, David J. (David John)
 Bridge bearings and expansion joints/David J. Lee. – 1st ed.
 p. cm.
 Includes bibliographical references and index.
 ISBN 0–419–14570–2 (alk. paper)
 1. Bridges–Bearings. 2. Bridges–Floors–Joints. I. Title.
TG326.L42 1994
624'.257–dc20

93-32192
CIP

Contents

5   Applications   139

Introduction   139

Preface

This volume is a revised, expanded and updated version of the present author's *The Theory and Practice of Bearings and Expansion Joints for Bridges* published by the Cement and Concrete Association in 1971. The original book attempted to address the introduction at that time of new types of bearings and expansion joints and the consequent inter-relationship with the practice of bridge design. Over 20 years have elapsed, enabling the lessons of experience to be applied, improvement and consolidation of the principles evolved, and the underlining of the virtues of high quality initial design and installation for achieving good durability and serviceability.

The publication of BS 5400: Part 9 in 1983 made available for the first time in the UK a British Standard dealing specifically with bridge bearings. This present volume therefore takes into account the design rules contained in Part 9.1, which is the Code of Practice and the specification requirements in Part 9.2 for materials, manufacture and installation of bearings.

The book is written primarily as an aid to the practising bridge engineer in seeking and fulfilling the objectives of sound design. Increased importance is now given to design which ensures accessibility for inspection of bridge components and ease of maintenance. Also some advice is proffered relating to bridge refurbishment and strengthening.

Bridge engineering is an art as well as a science and sometimes clients force a compromise of the two. However, works of art are not like curates' eggs, only good in parts. Nothing inspires the engineer or layman alike more than a daring bridge design, but what makes a really good bridge is how the structure manages to integrate all the demanding and conflicting requirements over a period of time and still give artistic satisfaction.

David J. Lee

Acknowledgements

I am grateful to the manufacturers of bearings and expansion joints who have readily supplied many of the illustrations. These are acknowledged appropriately with the illustration, as are any photographs from other sources. Where an acknowledgement is not given it has usually been obtained from the photographic library of the Maunsell Group.

I would like to thank in particular two colleagues at G. Maunsell & Partners, Ron Bristow and Alec Wallace, who have been liberal with their extensive experience and wide knowledge. Most important to me has been their encouragement in completing a task that, though esoteric, is highly important for bridge engineers.

To Alec Wallace with Kind Regards

David Lee

ONE

Movements

1.1 Historical background

Structural engineering in the past had to grapple with the problem of foundation stability to a far greater extent than the difficulty of accommodating thermal strains in the structure once it was standing; it is largely the modern requirements for bridge structures which have generated the problems of bearings and expansion joints. Not only are spans larger, but materials such as steel and concrete are manufactured as monolithic materials. The contrast between these and the fissured and jointed materials like stone and timber is striking. History, then, does not provide much background to what is essentially a modern problem.

The stone lintel, with its necessarily limited span, laid down shear strength and bending stress as the criteria of design. Movement of the beam under load and temperature change were of no practical significance. When spans were increased with the invention of arch construction they had to be made of relatively small pieces. In consequence, thermal movement took place by a geometrical change of shape in the structure, with the small displacements allowed for by dry jointing the stones of the arch and loose filling in the spandrels. When one thinks that such principles were in use for arch construction right up to the nineteenth century, it is realized that the problem of the mechanical engineering treatments required for movement is relatively new.

During the boom periods of canal and railway construction, many multispan brick and masonry viaducts were built. These contain no expansion joints but generally have shown only limited signs of distress. In some instances it has been established that the temperature changes within these relatively massive structures are substantial but that the resulting expansion or contraction has been accommodated more by elastic strain than by hogging or sagging of the arch spans [1]. More recently, several multispan structures have been designed to take advantage of the prestressing forces which are generated by this arrangement. These can be beneficial for internal spans but require special attention to be paid to the end 'anchor' spans.

Timber trusses (Fig. 1.1) can avoid the problem by virtue of the nature of the material. Either the truss can be pin-jointed and the temperature strains taken up with rotation at the joints or, as in some of the medieval hammer-beam roof trusses, complete stiffness can be built into the joints and movements accommodated by flexural distortion.

When concrete began to be used for arches, the brittleness of its monolithic character soon exposed the problem. In Glenfinnan Viaduct, Scotland, the first concrete railway bridge in Britain, 21 arches of 15 m span are formed of mass concrete (Fig. 1.2). The possibility of differential settlement of the piers was dealt with by incorporating a sliding joint at the crown of each arch consisting of two 13 mm thick steel plates placed together [2]. Since the viaduct was opened in 1898 there has been no noticeable movement of the piers, and the horizontally curving structure is still in good condition.

It is notable that in the first major use of structural steelwork in the Forth Railway Bridge, completed in 1890, the need to provide for movements was recognized. At the ends of the suspended spans, temperature movements of up to 200 mm were allowed for, and at the base of each cantilever tower, three of the four skewbacks were free to move to allow elastic shortening of the base struts between them [3].

Fig. 1.1 Brunel's timber Old Liskeard Viaduct, Cornwall. (Photograph taken 1894, reproduced by courtesy of British Railways, Western Region.)

Fig. 1.2 Glenfinnan Viaduct, Scotland. (Courtesy of British Cement Association.)

With highway bridges, not only has the intensity of loading tended to increase, but so also have the velocity and frequency of traffic to such an extent that fatigue and dynamic loading are subjects of concern [4] and will become increasingly so as the strength/weight ratio of steel or concrete is improved.

In railway bridges the development and wide adoption of continuously welded rails has required the effects of differential movements between the trackbed and the structure to be considered [5]. The rail expansion joints do not generally coincide with the bridge joints, and the degree of interaction will be different for, say, ballasted and plinth track. Deck drainage at the bridge joint requires special attention to avoid damage to the substructure.

All these factors mean that modern bridges are more sensitive to the movements that arise in them, and the generation and magnitude of these movements are of increasing importance.

1.2   Sources of movement in structures

An essential first step in assessing the problems arising from movement in structures is to set out the sources from which movements are likely to derive in any given situation and to set down quantitative data which will fulfil the resulting requirements.

One might think first of the distortions caused in a structure as a result of the physical properties of the materials used. These could be due to:
1. temperature and humidity changes;
2. creep, shrinkage and fatigue effects;
3. axial and flexural strains arising from dead and live loading, prestressing etc.;
4. dynamic load effects including impact, braking and lurching, centrifugal forces ;
5. overload.

Next one may set down the exterior sources of movement, such as:
1. tilt, settlement or movement of ground;
2. mining subsistence;
3. seismic disturbances;
4. moving parts of structures, e.g. in lift spans and swing bridges;
5. erection procedures.

The directions of movement, periods of application, whether they are reversible or not, and what flexural rotations are associated with components of the movement must all be determined.

For the design of a prestressed concrete box beam spanning approximately 60 m in an area of mining subsidence, it is of interest to record the movements in the longitudinal direction which were taken into account:
1. temperature   ±13 mm;
2. humidity   ±3 mm;
3. residual creep and shrinkage   −25 mm;
4. elastic prestressing deformation of concrete   −13 mm
5. 0.2% ground strain owing to mining subsidence   ±150 mm

These have been quoted for a particular instance to show how the various figures can add up.

The effects of thermal changes become progressively more important for large spans and continuous bridges, and are dealt with in some detail in this volume as recent investigations have given valuable data on which to base design criteria.

The question of whether movements are related to long- or short-term effects requires careful consideration, and the economic use of prestressing has generated an interest in the calculation of losses arising from change of strain with time in concrete under stress. The introduction of advanced composite materials into structures will require further consideration of time-dependent strain effects. In structural steelwork, thermal movements occur axially and flexurally, and even out of plane 'quilting' of the web plates of a box girder due to differential temperatures between the skin plating and internal stiffeners has been recorded [6]. Temperature gradients require particular consideration, e.g. in design of the shear connection in composite steel/concrete members. Guidance on most of these effects is given in BS 5400.

For special structures, such as large arches like Gladesville Bridge [7] and cable-stayed bridges, studies of the behaviour of the structure under load have led to a valuable extension of knowledge.

1.3 Temperature effects

1.3.1 UNIFORM CHANGE OF TEMPERATURE

In general, it is not desirable to restrain a structure fully against temperature movements. This becomes obvious if we consider a prismatic concrete beam of cross-sectional area 1.00 m^2. If this beam heats up by $10°C$ and is restrained from expanding axially, a stress of 3.60 N/mm^2 is induced and the restraining force required is 3.60 MN (361 tonf). These figures are based on a coefficient of linear expansion of $12.0 \times 10^{-6}/°C$ and a Young's modulus of $30\,000 \text{ N/mm}^2$.

However, in practice, some restraint is inevitable since completely frictionless bearings and rigid substructures do not exist. Differences in the expansion per unit temperature change due to the different restraint forces applied to end and centre spans have been recorded [8]. In order to estimate the effect of restraints, it is necessary to know what movements would take place if the structure were totally unrestrained. The actual movements of the structure can then be found by adding the movements caused by the restraining forces. Methods of estimating the restraining forces are discussed below.

It is well known that if a monolithic body heats up evenly, each part of the body will move along a line radiating from the point of zero movement. Also, the movement is proportional to the distance from the point of zero movement and to the temperature change. Hence, if we know the geometry of the body, the coefficient of linear expansion and the temperature change, it is easy to calculate the free thermal movement.

Typical values for the coefficient of linear expansion of various materials are given in Table 1.1 from which the importance of correctly identifying them is immediately apparent. This is particularly true when differential movements between them have to be considered.

1.3.2 DIFFERENTIAL TEMPERATURE EFFECT

Another effect is thermal bending, which can occur when different elements of a structure are at different temperatures (Fig. 1.3). The temperature distribution across a cross-section will generally not be linear

Table 1.1 Coefficients of linear thermal expansion

Material	Coefficient × 10⁻⁶ per °C	Reference
Metals		
Structural steel	12	BS 5400
Corrosion resisting steel	10.8	Cromweld 3CR12
Stainless steel: austenitic	18	BRE Digest 228 [9]
Stainless steel: ferritic	10	BRE Digest 228 [9]
White cast iron	8–9	
Grey cast iron	10–11	
Ductile cast iron	11	
Cast carbon steel	11.5–12.5	
Cast alloy steel	11.0–18.0	
Structural aluminium alloy	22–24.5	CP 118 [10]
Bronze	20	BRE Digest 228 [9]
Aluminium bronze	18	BRE Digest 228 [9]
Steel wire rope	12.5	British Ropes Ltd.
Concrete		
Gravel aggregate	13	Hammersmith Flyover
Limestone aggregate	8	Mancunian Way
Normal weight concrete	12	BS 5400
Lightweight concrete	7	
Limestone aggregate	9	
Chert aggregate	13.5	BD/14/82 [11]
Quartzite aggregate	12	
Sandstone, quartz, glacial gravel	11.5	
Siliceous limestone	11	
Granite, dolerite, basalt	10	
Limestone	9	
Cementitious mortar	10–13	BRE Digest 228 [9]

Material	Coefficient × 10⁻⁶ per °C	Reference
Masonry		
Concrete brickwork and blockwork		
Dense aggregate	6–12	BRE Digest 228 [9]
Lightweight aggregate (autoclaved)	8–12	
Aerated (autoclaved)	8	
Calcium silicate brickwork	8–14	
Clay or shale brickwork or blockwork	5–8	
Granite	8–10	
Limestone	3–4	
Marble	4–6	
Sandstone	7–12	
Slate	9–11	

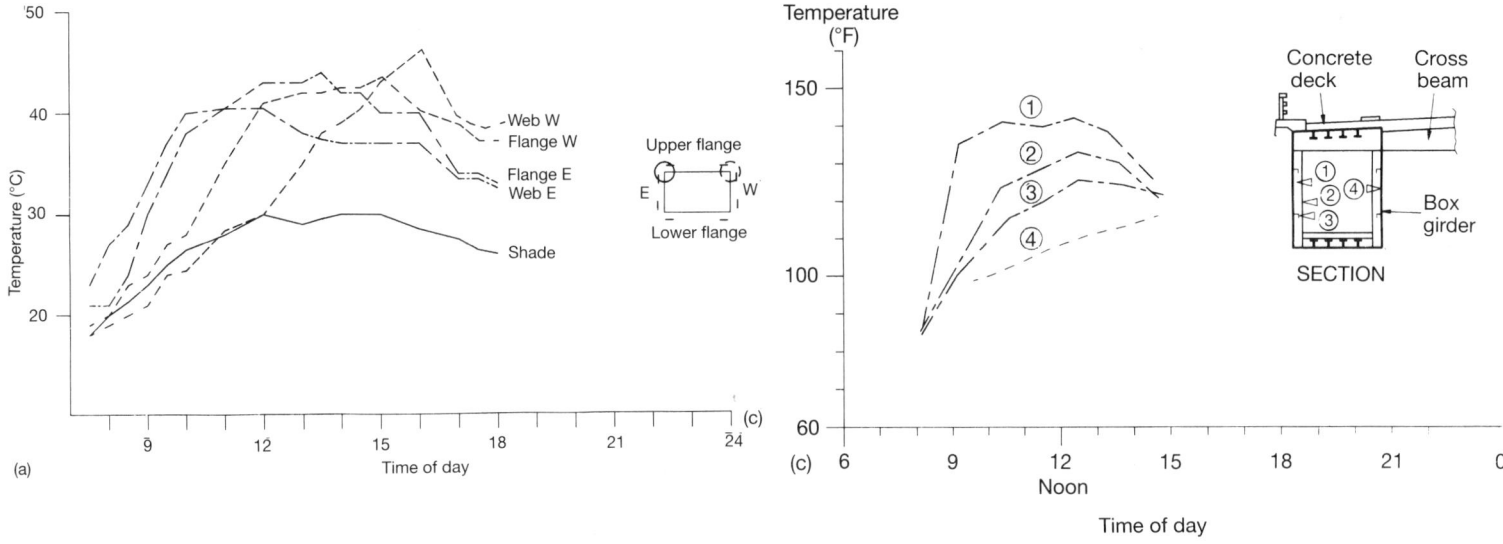

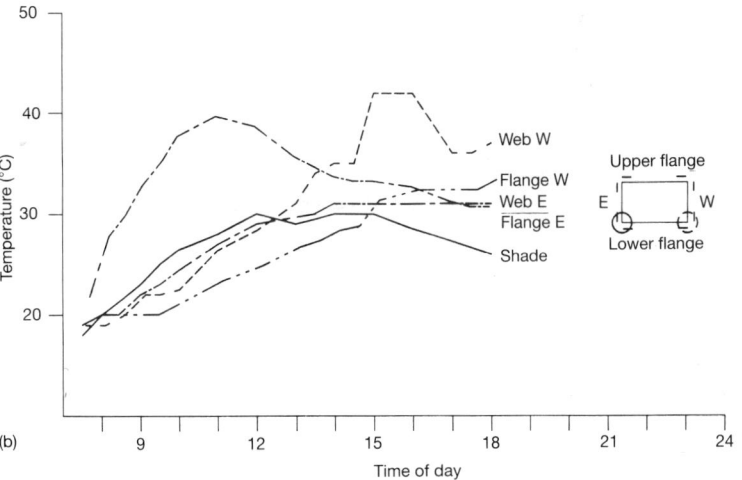

Fig. 1.3 Differential temperatures at (a) upper corners and (b) lower corners of a steel box girder during assembly in Iraq; (c) within a box girder web panel in Malaysia.

vertically, and also in some situations horizontally, hence an internal system of bending stresses will be induced such that the variation of elastic strain across the section is linear. This will, of course, only apply if we accept that plane sections remain plane, which is a usual design assumption. Although the induced stresses are of importance in the design of the section, as far as the movements produced are concerned, the only effect is that the structure takes up a curve of constant radius $R = D/\Delta$, where D is the overall depth of section and Δ is the difference in strain between extreme fibres. The curvature may be vertical, horizontal or a combination of the two. It can be seen from this that in order to calculate all the 'free' thermal movements we must know not only the coefficient of linear expansion of the material, but also the range of mean structure temperature that can occur and the worst possible temperature distribution.

For UK conditions, isothermal maps of minimum and maximum shade air temperatures derived from Meteorological Office data are incorporated in BS 5400 along with tables for minimum and maximum effective bridge temperatures [12]. These are based on the research programmes of the Transport and Road Research Laboratory which included measurements of movements on a number of major structures throughout Britain over a period of years, followed by theoretical studies which culminated in the methods of calculating the temperature distribution in bridge decks now incorporated in BS 5400 [13]. The results of this work are contained in various reports, some published by the Department of the Environment [14–19].

However, particular locations may still require special consideration. For example, wind chill and the capacity of the bridge to retain heat due to the differing materials used in various parts of the structure were reported as causing surface freezing on Foyle Bridge, Northern Ireland, when there were no reports of frost elsewhere in the area [20] and an ice detection system was installed (Fig. 1.4).

It is to be noted that the effective temperature of a bridge is defined as the temperature which governs longitudinal movement of the deck at the position of the neutral axis. Both thermal bending and lateral movement have a second order effect on the calculation of this temperature ([13], Appendix 1).

Prediction of the temperature distribution through sections requires a knowledge of heat transfer coefficients, intensity of solar radiation and shade temperatures. The values of heat-transfer coefficients have to be estimated for the particular materials being used, values for both concrete and steel surfaces being quoted in references [14] and [15] which, when used for particular bridges, gave good agreement with measured temperatures. The intensity of solar radiation is dependent on the latitude in which the structure is situated, its orientation, the season of the year and the hour of the day. The shade temperature is also dependent on the location of the structure and, as with solar radiation, reference must be made to local meteorological records. Once these factors are known, the temperature distribution and hence the mean temperature can be determined for any cross-section.

Fortunately, however, the reports come to some simplified conclusions regarding the range of mean bridge temperatures. These are as follows:

1. For concrete bridges, both the experimental and theoretical results suggest that the range of extreme mean bridge temperature is much the same as the range of extreme shade temperature in the same area.
2. For steel box-section bridges, both the experimental and theoretical results suggest that on cold winter days, the minimum mean bridge temperature will fall to 3 to 4 deg C below the minimum shade temperature; on hot summer days, the maximum mean bridge temperature will be about 1.5 times the corresponding maximum shade temperature in degrees Celsius, [21].

For the purpose of setting joints and bearings in the UK, the times when the effective bridge temperature is approximately equal to the mean shade temperature for that day are:

1. for a concrete bridge: 0800 ± 1 h;
2. for a steel bridge: 0600 ± 1 h [22].

There is, however, no simple way to determine the temperature distribution. For solid sections, an exponential distribution has been suggested which gives a reasonable simple result [23].

Fig. 1.4 Foyle Bridge, Northern Ireland. (Courtesy Ove Arup & Partners.)

For flanged sections we can make the simplifying assumption that the top and bottom flanges are at constant mean temperatures differing by T. This temperature distribution will give approximately the same effects as the true distribution, if the ratio of flange thickness to overall depth is small, with the exception that it underestimates the induced compressive stress in the extreme top fibre.

The method of calculation of vertical temperature distribution presented in BS 5400 (Fig. 1.5) and in the Transport and Road Research Laboratory (TRRL) reports [13,19,22] may be considered an appropriate basis for calculation of differential temperature in the UK and for deck members in which the slope never exceeds about 3°. However, for other climatic conditions, modifications may be required as can be seen in Figs 1.6 and 1.7 which show that for Hong Kong conditions, BS 5400 would underestimate the top surface temperature in a thick concrete slab or in a concrete box by about 4.5°C [24].

It is not usually necessary to consider horizontal curvature effects due to the transverse temperature distribution except in twin super-structures where one structure shades the other. This can cause the gap between them to open or close, and this movement must be combined with any longitudinal differential movement when designing a cover for the joint.

The redistribution of bearing reactions due to a combination of the vertical and horizontal thermal bending effects may be significant in curved or acutely skewed spans.

For tall vertical piers or the towers of cable-stayed and suspension bridges it will be necessary to modify some of the basic equations and the values of certain coefficients to allow for vertical surfaces and for the orientation of the structure.

For a steel structure the air inside a vertical box, in the absence of full diaphragms, may be free to flow upwards as it warms, and thus there is convection inside the surfaces of the box. The difference in temperature between the steel and the air inside the box will cause exchange of heat by radiation. The coefficients for heat transfer proposed in the TRRL reports [13,19,22] for horizontal structures do not allow for these effects and may require modification.

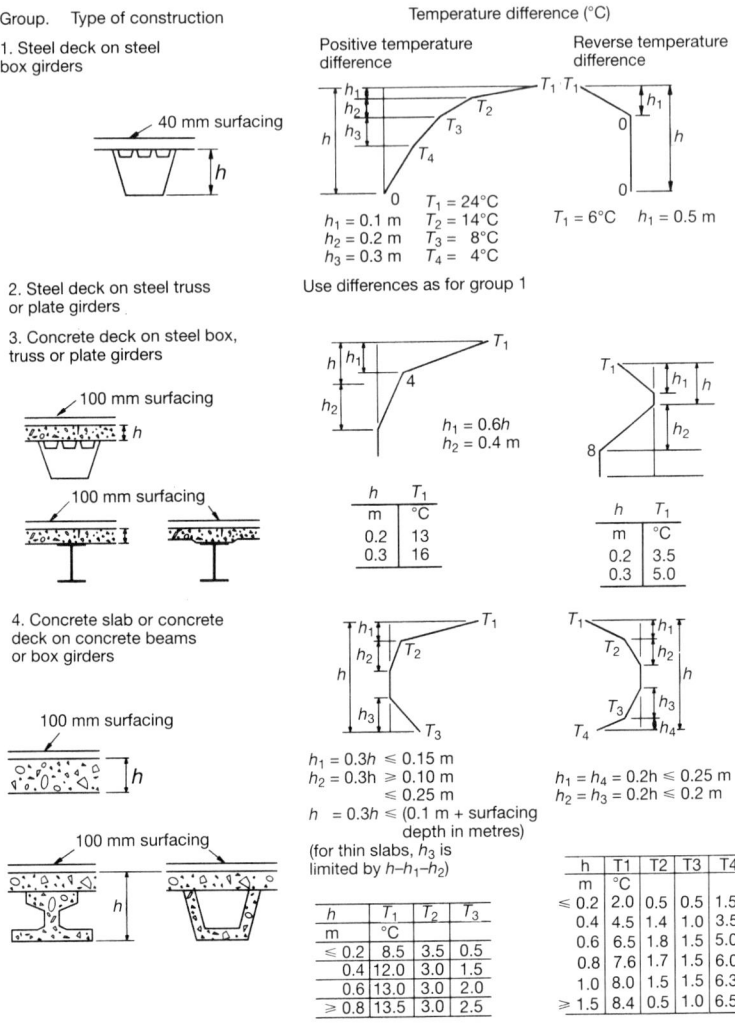

Fig. 1.5 Temperature difference for different types of construction. (Fig. 9, BS 5400: Part 2.)

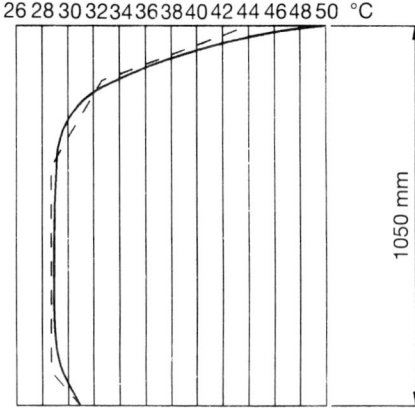

26 28 30 32 34 36 38 40 42 44 46 48 50 °C

1050 mm

Fig. 1.6   Comparison of positive temperature difference in a thick concrete slab calculated for Hong Kong (——) and by BS 5400 (– – –).

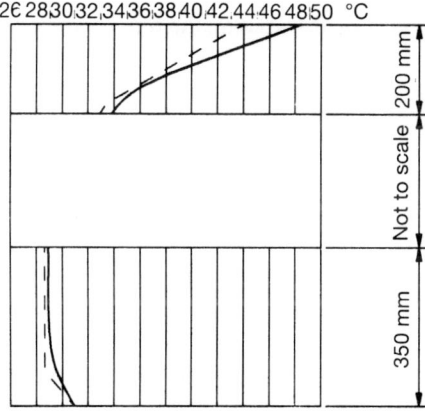

26 28 30 32 34 36 38 40 42 44 46 48 50 °C

200 mm

Not to scale

350 mm

Fig. 1.7   Comparison of positive temperature difference in a concrete box calculated for Hong Kong (——) and by BS 5400 (– – –).

It is also necessary to allow for the 'normal temperature' radiation from the external surfaces of the vertical box to the sky. This radiation is present at all times, but is small compared with the solar radiation on a horizontal surface. With vertical surfaces the solar radiation may be small. The 'normal temperature' radiation to the sky becomes more significant and allowance must be made. The basic equation therefore becomes:

Heat supplied by global radiation incident on steel less heat loss by radiation	Heat lost by convection and radiation at both surfaces of the steel	Heat into the steel to raise its temperature

$$(rI - 0.5eQ) \quad = \quad h_s \, (\Theta_s - \Theta_a) \quad + \quad \rho c x \, \frac{d\Theta_s}{dt}$$

where

r = absorption coefficient of external surface of steel (depends on the colour and type of paint);

I = intensity of incident global radiation (W/m²);

e = emissivity of surface = 0.9;

Q = 110 W/m² (0.5 eQ assumes that a vertical surface experiences one half of the radiation loss to the sky experienced by a horizontal surface);

h_s = combined heat transfer coefficient for both surfaces of the steel, taking into account both convection and radiation losses/gains (W/m²°C);

Θ_s = temperature of steel (°C);

Θ_a = shade air temperature (°C): range may be taken as 15°C;

ρ = density of steel (kg/m³);

c = specific heat of steel (J/kg°C);

x = thickness of steel (m);

$\dfrac{d\Theta_s}{dt}$ = rate of change of steel temperature with time (°C/s).

The values for global radiation may be obtained from the *Institution of Heating and Ventilating Engineers Guide, Book A, Design Data* [25] and other data on heat transfer coefficients for surfaces of different orientations and solar absorption coefficients from references given in report LR 561 [19]. In an unpublished study of the temperatures in the towers of the Forth Road Bridge the cross-section of which contains five cells it was proposed that, for evaluation of the 1 in 120 year maximum temperatures, the values of the heat transfer coefficients should be taken as tabulated below. In the same study, values of these coefficients were obtained indirectly by calculation from measured data and are shown in parentheses. It should be noted that lower values of h_s will give higher temperatures.

Face of tower	Proposed value h_s (W/m²°C)
North	19
South	13 (22 to 15)
East and west	16 (23 to 18)

The starting condition for calculation may be taken to be as assumed in the TRRL reports, but the time of starting should be modified to coincide with the time of zero radiation appropriate to the time of year being considered.

For a vertical concrete structure the method shown above for steel structures may be employed to obtain the corresponding modification to equation 5 in Appendix 1 of LR 561 [19], thus allowing the calculation method to be used to obtain the temperature distribution through a concrete pier or pylon. It will be necessary to modify the starting times and temperatures for the various areas of the vertical concrete structure.

1.3.3   RESTRAINTS ASSOCIATED WITH THERMAL BENDING

It has been shown above that, owing to differential temperature effects, a structure will want to take up a curve of radius R. If the structure is statically determinate, as in a simply supported beam or a cantilever, this can happen freely, producing end rotations.

In this context it is appropriate to point out that cantilevers are particularly sensitive to this sort of deflection. The leading cantilever of a continuous 61 m span concrete viaduct under construction in London was observed to deflect by 13 mm during the day and required a propping force of approximately 1 MN (100 tonf) to prevent this movement. It is also of interest to note that, during the construction of the Beachley Viaduct near the Severn Bridge (a steel box-girder deck), a temporarily cantilevered portion of deck was reported to oscillate perceptibly as clouds moved across the sun [14]. In tropical environments, horizontal and torsional movements occur during cantilever construction of steel box-girder bridges. These are particularly noticeable in the early sun and when the time available to secure the full perimeter of a box splice is limited [6].

In swing bridges differential temperature can cause high pressure on the nose wedge bearings making it difficult to withdraw them and open the bridge.

If, however, the structure is statically indeterminate, as in a continuous bridge, the redundancies will cause restraint. In bridge decks this problem can be solved by distributing the fixed end moments for each span. The fixed end moment M is independent of span length and is given by the formula

$$M = \frac{EI}{R}$$

where E = modulus of elasticity, I = moment of inertia and R = radius of curvature. The distributed moments produce rotations and deflections which can be calculated by means of standard methods. They also produce changes in the support reactions, which can be of significance especially if the spans are skewed.

A particular example for this occurs in the assembly of a number of steel girder sections on temporary supports to form a complete span. Differential temperature effects can cause a significant redistribution of reactions between the points of support which can result in any one or a combination of the following:

1. overload of temporary bearings;
2. differential settlements between the temporary supports;
3. difficulties in maintaining the unstressed camber;
4. difficulties in closing bolted joints or controlling welded joint gap settings prior to welding.

It has been necessary on some sites to incorporate special inter-linked supports which allow the vertical flexural movements to take place without changes in support reactions.

When assembling 14.5 m long steel panels to form sections of a box girder in Malaysia it was found on occasion that holes which had been reamed to match during shop assembly were as much as 6 mm out of register and could not be pinned and bolted until plate temperatures had balanced the following morning [6].

1.3.4 EXAMPLE SHOWING THE EFFECT OF THERMAL BENDING

In order to give some idea of the magnitude of the effects of thermal bending the following example (Fig. 1.8) shows that an overall posi-tive temperature difference of 11°C in the depth of the cross-section of a two-span concrete box beam bridge will change the self-weight reac-tion at the supports by as much as 13.4%.

The properties of the cross-section are:

area (A)	6.18 m²;
moment of inertia (I)	1.82 m⁴;
self-weight (ω)	14 850 kg/m;
coefficient of expansion (α)	12 × 10⁻⁶/°C;
modulus of elasticity (E)	34 500 N/mm².

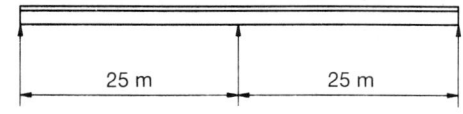

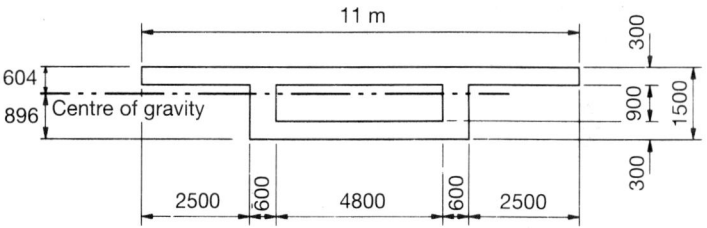

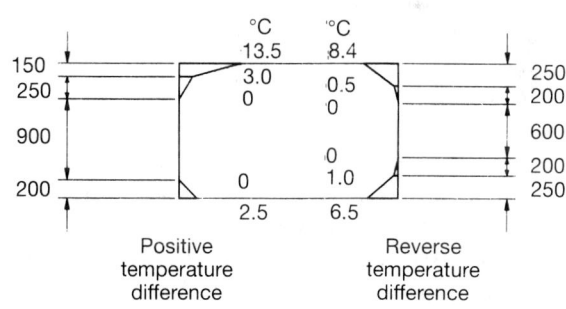

Fig. 1.8 Example showing the effect of thermal bending.

The fixed end moments applied to each span by the positive temperature difference where t is temperature of element, a is area of element and y is distance of element from neutral axis:

$$\sum E\alpha tay = \frac{34500 \times 12 \times 10^6}{10^3} [(3 \times 0.15 \times 11 \times 0.529)$$
$$+ (10.5 \times 0.15 \times 11 \times 0.554)/2 + (1.2 \times 0.15 \times 11 \times 0.379)$$
$$+ (1.8 \times 0.15 \times 11 \times 0.404)/2 + (1.2 \times 0.1 \times 2 \times 0.6 \times 0.271)/2$$
$$- (2.5 \times 0.2 \times 6 \times 0.829)/2] = 3123 \text{ kN m}$$

The resulting moment at the centre support is

$1.5 \times 3123 = 4685 \text{ kNm}$

Thus the end support reaction due to temperature difference is

$4685/25 = 187 \text{ kN}$

The end support reaction due to self-weight is

$0.375 \times 148.5 \times 25 = 1392 \text{ kN}$

Therefore the percentage change due to the positive temperature difference is

$187 \times 100/1392 = +13.4$

Similarly the percentage change in the central support reaction due to the positive temperature difference = -7.9.

The rotation at the end support due to the positive temperature difference is

$$\frac{3123 \times 25}{3 \times 2 \times 34.5 \times 1.82 \times 10^6} = 0.000207 \text{ rad}$$

1.4   Creep and shrinkage

Creep and shrinkage of concrete are both dependent on the actual materials used and only crude accuracy can normally be expected from the variety of creep and shrinkage coefficients which appear in different codes. These are primarily intended for use in the design of the structure rather than for estimating moments in joints and bearings where time dependency is of importance, and it is appropriate to be aware of the diversity of theories as no single theory is universally accepted. In addition to the movements which take place, the redistribution of bending moments as a result of creep in continuous construction changes the distribution of loads on bearings. A comprehensive review is available in specialist publications [26].

The mean values given in Table 1.2 may be adopted for the final shrinkage and for the final creep coefficient for a concrete subjected to a stress not exceeding $0.4f_{cj}$ at age j from loading. They apply to concrete of medium consistency, made with rapid hardening cement and for constant thermo-hygrometric conditions (the mean temperature of the concrete being 20°C and the relative humidity as indicated).

The development of concrete deformation with time can be estimated from Fig. 1.9.

In simple cases, in particular in reinforced concrete, the effects of creep may be estimated by reducing the modulus of elasticity E_{cm} to one-third of the values given in Fig. 1.10.

In special cases, such as free cantilevering construction or structures where concretes of different ages are combined, more detailed information is needed and reference should be made to the CEB *Design Manual on Structural Effects of Time Dependent Behavior of Concrete* [26] and the CEB *Manual of Cracking and Deformations* [27]. For a rigorous assessment of temperature effects in complex concrete structures refer to the UEG Report, 'Designing for temperature effects in concrete offshore oil-containing structures' [37].

In order to explain in real terms what the basic theories imply, the example of two spans of beams, erected as simply supported beams and made continuous, will be used (Fig. 1.11). This example is chosen as it features in many Codes of Practice. By simplifying the calculations, the

Table 1.2 Final values for shrinkage and creep coefficients for concrete

	Atmospheric conditions			
	Humid[1]		Dry[2]	
	Equivalent thickness[3]		Equivalent thickness[3]	
	Small	Large	Small	Large
Shrinkage[4]				
Fresh (3–7 days)	0.26	0.21	0.43	0.31
Medium (7–60 days)	0.23	0.21	0.32	0.30
Mature (> 60 days)	0.16	0.20	0.19	0.28
Creep[5]				
Fresh (3–7 days)	2.7	2.1	3.8	2.9
Medium (7–60 days)	2.2	1.9	3.0	2.5
Mature (> 60 days)	1.4	1.7	1.7	2.0

[1]Relative humidity approx. 75%.
[2]Relative humidity approx. 55%.
[3]$2A_c/u$ (small: $\leqslant$ 200 mm; large: $\geqslant$ 600 mm), where A_c denotes the area of the concrete section, and u is the perimeter in contact with atmosphere (which includes the interior perimeter of a hollow section only, if there is a connection between the interior and the free atmosphere).
[4]$\varepsilon_{cs}(t_\infty, t_0) \times 10^3$, where t_0 is the age of the concrete at the instant from which the shrinkage effect is being considered.
[5]$\phi(t_\infty, t_0)$, where t_0 is the age of commencement of loading.

design engineer can better appreciate what is happening to the structure and through this understanding guard against errors which might otherwise pass unnoticed.

The support moment initially is zero but due to creep a moment develops with time. This moment (M_∞) may be expressed in terms of the moment (M_1) which would have occurred if the structure had been built as one, e.g.

$$M_1 = \frac{wl^2}{8}$$

for a uniform section beam with self-weight w per unit length.

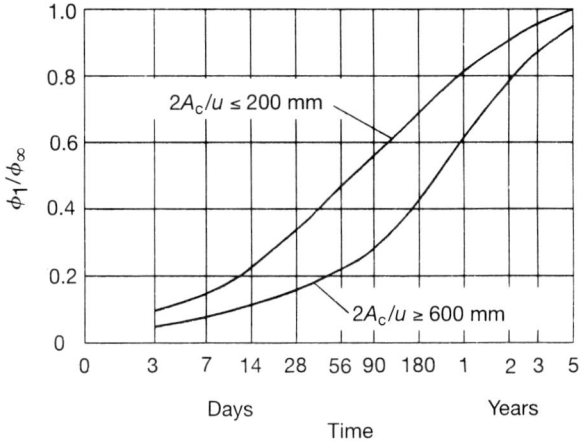

Fig. 1.9 Variation of creep coefficient with age.

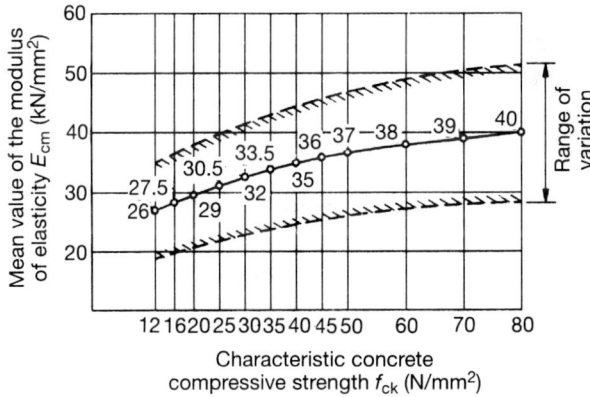

Fig. 1.10 Variation of modulus of elasticity with compressive strength.

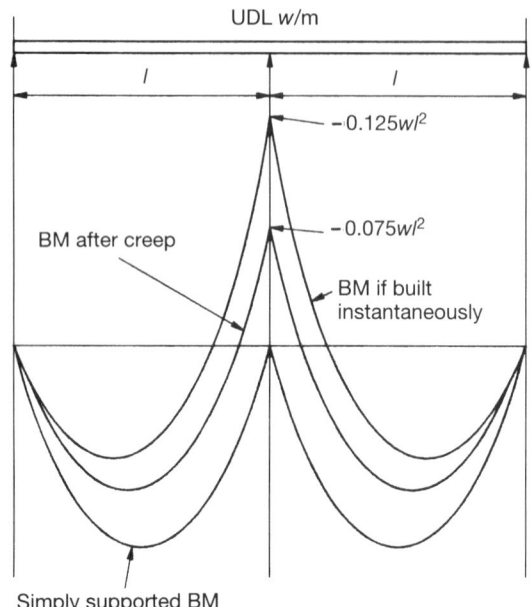

UDL *w*/m

l   *l*

−0.125*wl*²

BM after creep

−0.075*wl*²

BM if built
instantaneously

Simply supported BM

Fig. 1.11 Bending moments (BM) in precast beams made continuous after erection.

The value of the final support moment is conveniently expressed as follows:

$$M_\infty = M_1 K f(\phi)$$

where K is a factor depending on the age of the beams when they are made continuous, and ϕ is the creep coefficient. This creep coefficient is a measure of the creep strain resulting from unit stress such that

$$\text{Creep strain} = \frac{\phi}{E_c}$$

where E_c is the modulus of elasticity of the concrete. The function $f(\phi)$ is not absolutely defined and there are a number of different formulae

that have been proposed for it. The simplest of these formulae is based on the 'effective modulus' method and gives

$$f(\phi) = \frac{\phi}{1+\phi}$$

This formula is given in the FIP Recommendations [28]. In the specialist literature on creep, it is recognized as being very much a simplification of reality and as such is not likely to be particularly accurate.

Another formula is based on the 'rate of creep' method and gives

$$f(\phi) = (1 - e^{-\phi})$$

This formula is given in BS 5400: Part 4 [29]. It has some attraction as it is mathematically elegant. However, it is also based on some simplifying assumptions and should not be regarded as being completely accurate.

A third formula based on the 'Trost–Bazant' method [30] gives

$$f(\phi) = \frac{\phi}{1+X\phi}$$

where X is termed the 'ageing coefficient' and is similar to the 'relaxation coefficient' η given in Appendix C to BS 5400: Part 4 [29]. For practical purposes, a value of $X = 0.8$ can be adopted.

Values of ϕ are generally in the range 1.5 to 2.5 and Fig. 1.12 compares the different values of $f(\phi)$ for this range.

It is tempting to regard the first two formulae as being lower and upper bounds and the third as being a more realistic compromise. However, in reviewing the various methods for predicting creep and its effects Neville *et al.* [30] state.

> ... it is clear that, at the present time, no method of analysis and no creep function can be considered to be exact.

Studies which have attempted to measure the induced moments [31] have concluded that it is not possible to confirm the accuracy of any one formula. It therefore appears that the level of inaccuracy is such that it is reasonable to use any of the formulae. It is, however, important to bear in mind that results can vary and one should design for the worst case.

Generally, Codes of Practice take no account of the time that the beams act simply supported before they are made continuous. This has a significant effect on the moments induced as creep takes place relatively quickly in the early life of the beams.

The means of taking this into account is to include the factor K in the function defining M_∞. This factor is

$$K = \frac{\phi_\infty - \phi_t}{\phi_\infty}$$

where ϕ_∞ is the creep coefficient relevant to infinite time, commonly abbreviated to ϕ.

ϕ_t is the creep coefficient relevant to time t, where t is the time from casting of the beams to when they are made continuous.

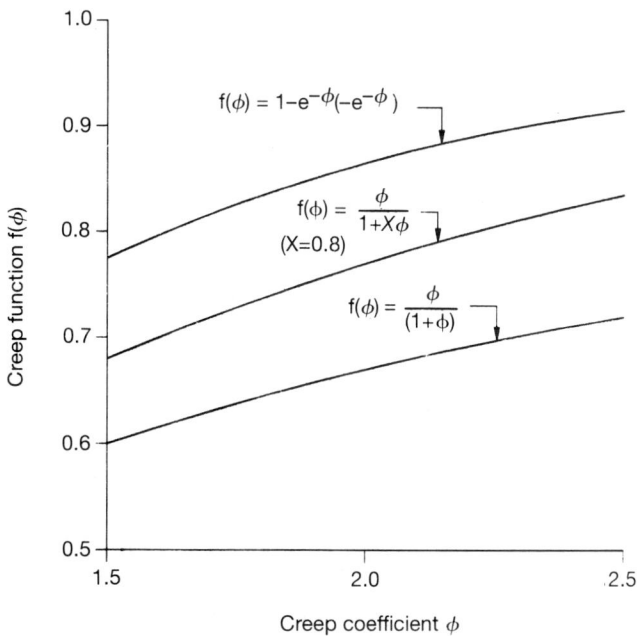

Fig. 1.12   Comparison of creep functions.

It may be noted that using the notation of BS 5400: Part 4: Appendix C,

$$K = \frac{k_j - k_{jt}}{k_j}$$

where coefficient k_j defines the development of the time-dependent deformation with time and k_{jt} defines k_j at time t.

Also as $k_j = 1.0$,

$$K = 1 - k_{jt}$$

Hence, for example, a beam with an effective thickness of 300 mm made continuous 30 days after initial loading has $k_{jt} = 0.3$ and $K = 0.7$.

For $\phi = 2.0$ and using $f(\phi) = 1 - e^{-\phi}$

$$\text{Support moment} = \frac{wl^2}{8} \times 0.7 \times 0.86$$

$$= 0.075\, wl^2$$

The span-by-span method of construction involves the progressive construction of a multispan continuous structure in stages. For convenience of coupling prestressing tendons the joints between stages are spaced at about 0.2 of the span length.

The load applied in one stage creates bending moments in the structure which is completed at that stage. As each new stage is added, the load from that stage modifies the moments in the previous stages. Also the addition of each new stage provides an additional constraint on the previous stages and their moments are modified by creep. As with the example of the simply supported beams made continuous, the moments tend towards the moments that would have occurred in the final structure. However, for any one stage the structure itself is extended in stages. It is, therefore, necessary to consider the moments that the load of each stage would induce not only on its initial structural configuration but also on all subsequent structural configurations [38]. For the case of an internal span of a long viaduct of uniform section with joints at the 0.2 span length positions, these moments are shown in Fig. 1.13 and the support moments are shown in Table 1.3.

Table 1.3 Moments due to stage N load applied to various stages of structural completion[1]

Last stage built	N-6	N-5	N-4	N-3	N-2	N-1	N	N + 1	N + 2	N + 3	N + 4	Σ N
					Support							
N	-0.0001	0.0003	-0.0010	0.0038	-0.0142	0.0530	0.0200					0.0618
N + 1	-0.0001	0.0002	-0.0008	0.0031	-0.0117	0.0436	0.0548	0				0.0891
N + 2	-0.0001	0.0002	-0.0008	0.0031	-0.0115	0.0429	0.0573	-0.0094	0			0.0817
N + 3	-0.0001	0.0002	-0.0008	0.0031	-0.0115	0.0429	0.0575	-0.0101	0.0025	0		0.0637
N + 4	-0.0001	0.0002	-0.0008	0.0031	-0.0115	0.0429	0.0575	-0.0101	0.0027	-0.0007	0	0.0832

[1]All moments times wl^2. Cantilever length = 0.2l.
For a more general treatment see [38].

The final moment at a pier is

$$M_N + (M_{N+1} - M_N)K_1 \mathrm{f}(\phi) + (M_{N+2} - M_{N+1})\, K_2 \mathrm{f}(\phi) + \ldots$$

Assuming a 30 day continuous cycle per span and $\phi = 2$, the final support moment for the full dead load is in the range 0.075 wl^2 to 0.072 wl^2, depending on which $\mathrm{f}(\phi)$ is used. It is interesting to note that this range of moments is very narrow, deviating only ±2% from the mean. This shows that the relatively large variations in the function $\mathrm{f}(\phi)$ have little effect on the end result, and in these circumstances it is not important which function is adopted.

The above discussion has been related to the self-weight moments in the structure. Secondary prestressing moments are subject to similar effects and can be treated in the same way.

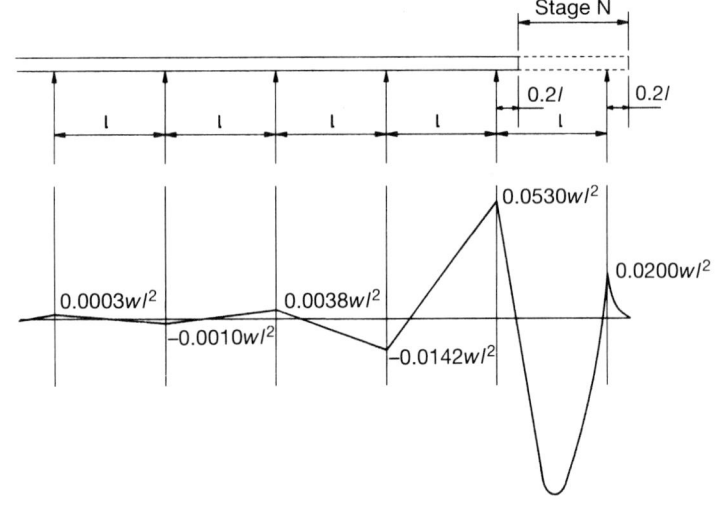

Fig. 1.13 Multispan viaduct – moments due to stage N self-weight.

1.5 Axial and flexural strains

The movements arising from axial strain are readily visualized. However, those resulting from flexural strain are less obvious as they arise from the rotation of the member and the distance from the neutral axis to the surface being considered (Fig. 1.14). As a result this movement is generally greater on the sliding surface of a bearing than in the road joint adjacent to it. In a flexible girder the number of longitudinal bearing movements due to live load can be more than 10 times those due to temperature, and when aggregated become the dominant factor in determining the life of a PTFE sliding surface. It is worth noting that in a simply supported span fixed at one end the flexural movement at the free end is doubled (Fig. 1.15).

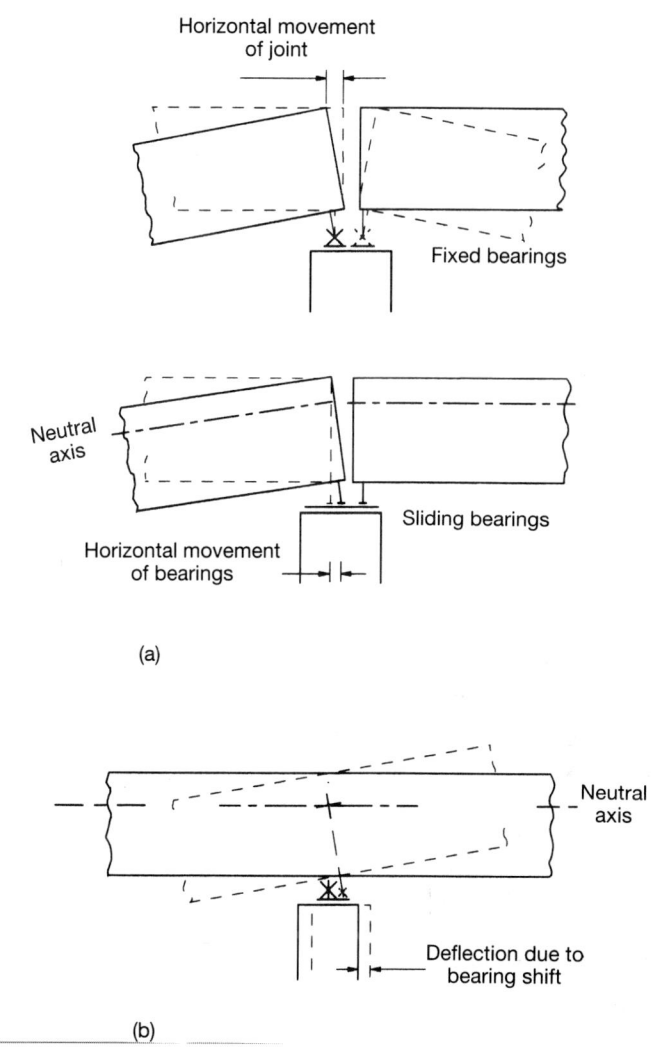

(a)

(b)

Fig. 1.14 Movements resulting from flexure of superstructure, (a) simply supported spans; (b) continuous spans.

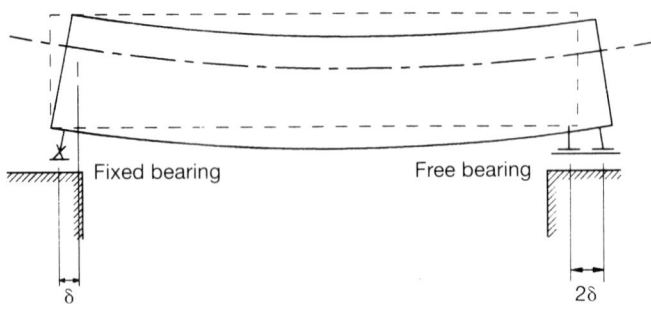

Fig. 1.15   Double movement at free end of simply supported span.

1.6   Dynamic loads

Rapidly applied loads cause movements of a structure which are usually oscillatory in character due to the elastic and internal damping properties of structural materials. They will therefore determine the maximum instantaneous movement for which provision is required and affect the rate of wear of moving parts in joints and bearings. These may, however, be alleviated by flexibility in the substructures. Additional external dampers can be introduced to reduce these effects in long slender spans or cantilevers, and shock transmission units (Fig. 1.16) used to distribute load over a number of spans [32].

1.6.1   MOVING STRUCTURAL ELEMENTS

Such elements are found, for example within lift spans or the complete superstructure in the case of a swing bridge. Limit switches control the acceleration/deceleration forces and buffer stops absorb the impact at the end of the move. However, provision for a degree of overrun is recommended and the elastic spring movements of bearings and their supports must be allowed for at joints. Roller bearings with a frictional resistance of less than 1.5% will generally be used.

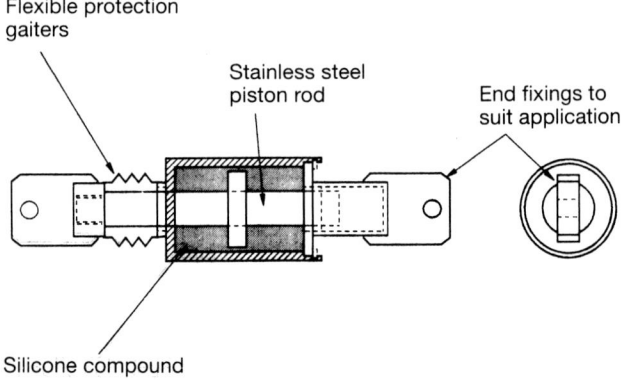

Fig. 1.16   Shock transmission unit.

1.6.2   LIVE LOADING

Within a span or series of continuous spans the rate of application of vertical live load is allowed for in design codes and no special provision need be made for it unless the structure is unusually slender.

For a cantilever, however, the load is applied instantaneously as a wheel comes onto or leaves it and the joints must be capable of accommodating the resulting dynamic movements. The possibility of dynamic uplift on bearings particularly at acutely skewed ends of spans also requires particular attention.

For longitudinal dynamically applied forces the spring stiffness of the substructures may be particularly significant when considering the movement joint requirements. Hydraulic creep couplers which allow thermal movements but resist suddenly applied loads may then be introduced to distribute the force over a number of spans [5]. These are particularly appropriate in the case of railway bridges in which the rail upon which the longitudinal force is applied is continuous over the joint between spans (Fig. 1.17).

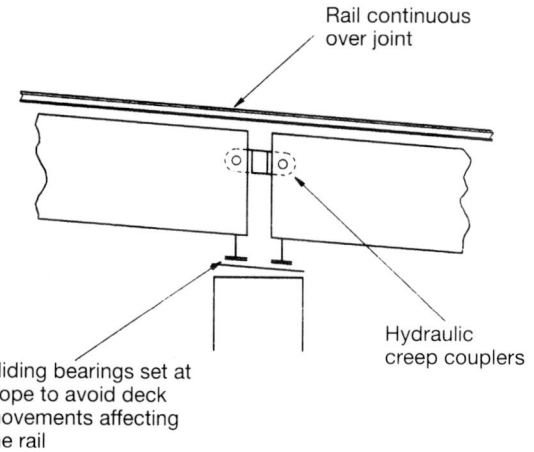

Rail continuous over joint

Sliding bearings set at slope to avoid deck movements affecting the rail

Hydraulic creep couplers

Fig. 1.17   Joint between simply supported railway bridge spans with continuous rails.

1.6.3   SEISMIC EFFECTS

Isoseismal maps are available for the UK and worldwide [33]. Further information can be obtained from the British Geological Survey and Central Electricity Generating Board (CEGB) studies for nuclear power stations. Known active areas include Colchester (where in 1884 an earthquake measuring VIII MM was recorded), Derby, the Great Glen from Inverness to Fort William and a diffuse area from South Wales and Herefordshire into the north Midlands [34]. Having assessed the level of seismicity the appropriate design earthquake is determined taking into account the probability of exceedence, risk assessment and reliability analyses [35]. The forces and movements on bearings may be distributed by the use of hydraulic creep couplers. The frictional resistance of PTFE sliding surfaces is also considerably higher under seismic movement than thermal or creep movement.

An alternative for structures in areas of high seismicity is to use simply supported spans which effectively float on elastomeric bearings incorporating a central lead plug damper.

1.6.4   ACCIDENTAL IMPACTS

Bearings are particularly vulnerable to damage as a result of 'bridge bashing' which continues to be a problem on road and river crossings.

The impacts on the superstructures may be accompanied by a significant uplift movement at the bearings. Where this is a possibility a positive tie down is required. For drop-in steel spans this has sometimes been provided by a link bearing. However, failure of such a link can result in overload of other adjacent bearing links, leading to a progressive failure.

Ship impacts more commonly occur against substructure caissons and piers and their effect on superstructure bearings are very dependent upon the relative flexibility and dynamic response of these components and of the foundation. Partial or complete freedom of movement on bearing surfaces should not be relied on for reducing the force transmitted to the superstructure as this requires a complete knowledge of initial bearing settings during construction, the position of the bearing due to temperature, the degree of slip/stick friction etc., at the moment of impact.

1.7 Overload

From time to time, bridge structures are checked for their capacity to carry a designated overload. The vertical load capacity of bearings will usually be checked for this, but movements at joints and bearings can also be greater than normal and must not be overlooked.

1.8 Foundations

Foundation movements, elastic or inelastic, will be reflected in the freedom required or forces generated at bearings and expansion joints. It is by no means unknown to find the girders of older bridges propping the abutments. In newer structures horizontal movements of recently compacted embankments can cause significant flexural movement of bankseat abutments, particularly if they have been constructed over soft strata, and the movement capacity of road joints in such a location must not be underestimated. Bearing movements may be affected by vertical settlement (uniform or differential), tilt or translation of substructures.

1.9 Construction

1.9.1 GENERAL

In multispan structures the movements at bearings and expansion joints can be significantly different during construction from those in the finished condition, and these must be allowed for to avoid the possibility of damage. Sources of excessive movements include:

1. thermal effects on steelwork exposed prior to placing deck concrete;
2. initial shrinkage of concrete;
3. elastic shortening due to prestressing;
4. temporarily changed positions of bearing restraints;
5. flexibility of temporary supports;
6. temporary immobilization of movement joint;
7. bearing rotation during cantilevering.

On Westway in London, a prestressed concrete elevated roundabout, which is approx 120 m internal diameter supported on 20 columns and connecting to several slip roads, required careful planning of supports during construction (Figs 1.18 and 1.19). A special sequence of *in-situ* casting and stressing followed the span-by-span method. As the first span cast could not be fully stressed until the ring was complete, the scaffolding supporting it was used as a rigid anchor during construction. To permit the movements due to thermal effects, concrete shrinkage and elastic shortening due to segmental prestressing, the bearings were fitted with temporary runner bars with a wide clearance which had to be calculated for each location. After completion of the ring these were removed, and purpose-measured permanent runner bars were made and fitted such that all subsequent movement was radial, the circumferential movements being restricted to approximately 6 mm [36, 38].

In cable-stayed and suspension bridges the cable extensions and sag changes which occur during construction as dead load stresses increase cause significant changes of geometry requiring the movement range of lateral bearings at the towers to be extended beyond the normal service range or by the provision of special temporary bearings.

In composite structures it is generally considered advisable to ignore concrete shrinkage when assessing the movements of expansion joints.

Bearings, when initially erected, will probably have temporary travel straps on the sides locking the upper and lower parts of the bearing together at a predetermined position. These should not be released until a sufficient load has been applied to prevent separation. Pot bearings are particularly sensitive to this and it is normally necessary to apply at least 25% of the vertical working loads to them to avoid potential problems, as any rotational movement will compress one side only and may cause separation at the other side, possibly resulting in damage to the rubber pad.

Initial bearing friction on PTFE can be high. Cases have been reported of cement or other fine dust getting onto contact surfaces as a result of the bearing having been opened up on site, causing even higher frictions. Although this did not happen on Westway it was noticed that some of the bearings did not move when first installed.

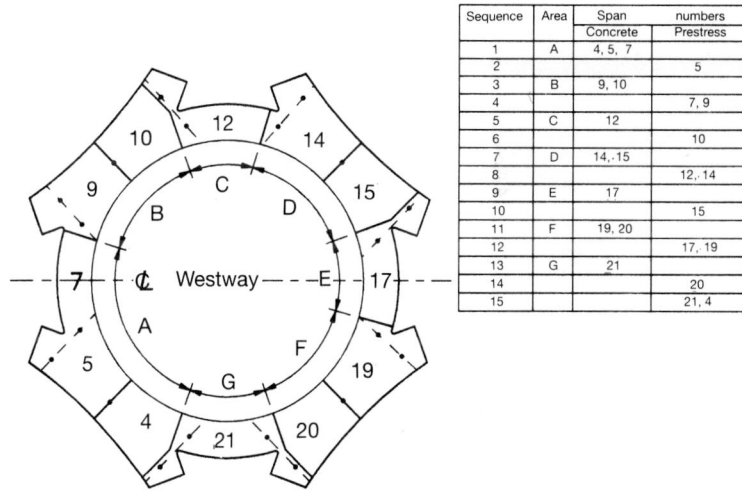

Sequence	Area	Span Concrete	numbers Prestress
1	A	4, 5, 7	
2			5
3	B	9, 10	
4			7, 9
5	C	12	
6			10
7	D	14, 15	
8			12, 14
9	E	17	
10			15
11	F	19, 20	
12			17, 19
13	G	21	
14			20
15			21, 4

Fig. 1.18 Construction of elevated roundabout in Westway.

Fig. 1.19 Westway, London (a) elevated roundabout complex looking west; (b) looking east towards Paddington.

b

They were then jacked back and forward a few times until they moved freely [36]. On a recent inspection after almost 20 years service they were still found to be functioning satisfactorily.

A brief mention might be made of some of the methods whereby controlled movements may be introduced in a structure by jacking during construction in order to obtain the required distribution of strains or stresses. After completion, the replacement of defective elements or access for maintenance and inspection purposes may also be required, involving further jacking, although it is generally desirable to keep such work to a minimum. It is often necessary to monitor clearances in joints during jacking to relieve a bearing of load. Sudden movements due to the release of locked-in friction forces must be anticipated. Stories abound of starting to jack without releasing bearing holding-down bolts.

When jacking one side of a twin superstructure deck, transverse horizontal movements have the effect of opening or closing the joint gap between them. This is particularly noticeable in curved and super-elevated structures.

1.9.2 HYDRAULIC JACKS

In many bridges, provision has to be made for jacking in order to modify the distribution of stresses or strains in the structure or to counteract subsidence. Similarly, the newer types of bearing generally bring with them the prudent requirement that provision has to be made for replacing them should they give trouble during the life of the bridge. Hydraulic jacking is the usual method by which this is achieved. Hydraulic jacks vary in capacity from very nominal values up to well over 1000 t with strokes from 10 to more than 500 mm. In all cases, care must be taken in the design of the structure to allow space for inserting a conventional jack; also, the reactions of the jack must be properly taken into account through the deck and substructure. Modern hose couplings are normally fitted with non-return valves, but the designer should not allow the situation where a burst hose or leaking hydraulic jack presents an unforeseen hazard. Locking, propping and wedging should be employed to prevent collapse or jamming of the jack through any failure. Similarly, it should be remembered that there may be a loss of pressure when the system stands for any appreciable period of time and due to settlement as packers, etc. settle down.

If a structure remains supported on jacks for more than a short period, allowance must be made for thermal movements over them to ensure that overloading does not occur and that they cannot be overturned.

1.9.3 FLAT JACKS

Flat jacks (Fig. 1.20) derive from the brilliantly inventive engineer Eugene Freyssinet. Full information about them is easily obtained from the manufacturers. A notable application in bridge construction was the jacking of the arch ribs of Gladesville Bridge in order to strike the centring (Fig. 1.21). Flat jacks are inexpensive, but it is even more important to provide a dual system so that, if one circuit fails, the load can be safely taken by the duplicate system. They can also be calibrated in advance and used at bearing locations to confirm the distribution of reactions.

1.9.4 SAND JACKS

These cheap devices, also referred to as sand boxes, have been found very satisfactory when controlled lowering is required, such as lowering beams on to their bearings or striking falsework. They consist of a piston set in a sand-filled cylindrical pot. In the simplest form the circular piston and base can be hardwood with the steel plate forming the cylinder walls designed on the basis of water pressure. Removal of the plug from a small side orifice, possible 25 mm diameter, at the base of the cylinder wall allows the sand to flow out under load in an easily controlled manner. Lowering is stopped by re-inserting the plug. Provided the sand is clean and maintained in a dry condition, jack performance is extremely reliable and it is not even necessary to provide a duplicate system. In the British climate, the sand is kept dry simply by covering the body of the jack with a polythene cover. Sand jacks were extensively used on Gladesville Bridge and on the Westway project.

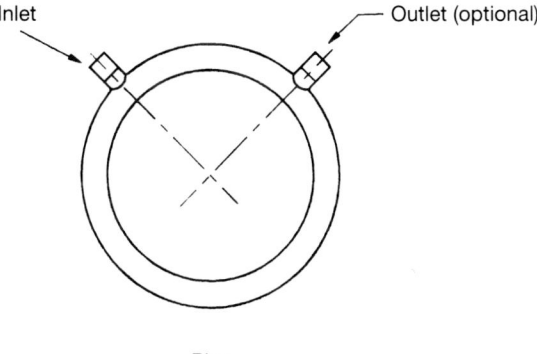

Inlet

Outlet (optional)

Plan

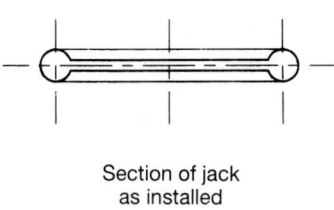

Section of jack
as installed

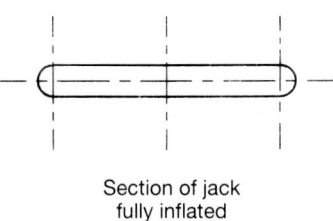

Section of jack
fully inflated

Fig. 1.20 Freyssi flat jack.

Fig. 1.21 (*Overleaf*) The Gladesville concrete arch ribs were made self-supporting off the falsework by the use of Freyssi flat jacks. (The completed bridge is shown in Fig. 5.16.)

References

1. MORRIS, E.H. (1948) The reconditioning of a defective arch in Stockport Viaduct. *Proceedings of the Institution of Civil Engineers*, **31**, 82–90.

2. WILSON, W.S. (1940) Advances in construction methods and equipment. *Proceedings of the Institution of Civil Engineers*, **13**, 273–304.

3. SHIPWAY, J.S. (1990) The Forth Railway Bridge Centenary 1980–1990 – some notes on its design. *Proceedings of the Institution of Civil Engineers*, **88**, 1079–1107.

4. British Standards Institution (1980) *BS 5400: Steel, Concrete and Composite Bridges: Part 10: Code of Practice for Fatigue.*

5. IABSE (1984) *International Association of Bridge and Structural Engineering, 12th Congress*, Vancouver, Canada, 3–7 September.

6. LEE, D.J. and WALLACE, A. (1977) Ahmad Shah Bridge, Malaysia. *Proceedings of the Institution of Civil Engineers*, **62**, 89–118.

7. FORRESTER, K. and CURTIS, T.J. (1967) Interim report on Gladesville Bridge arch deflections. *Institution of Engineers, Australia, Civil Engineering Transaction*s, **CE9**(2).

8. TANAKA, R., NATSUKAWA, K. and OHIRA, T. (1984) Thermal behaviour of multispan viaduct in frame. In *International Association of Bridge and Structural Engineering, 12th Congress*, Vancouver, Canada, 3–7 September.

9. Building Research Establishment (1979) Estimation of thermal and moisture movements and stresses; Part 2, Digest 228, Watford.

10. British Standards Institution (1969) The structural use of aluminium, Code of Practice CP118.

11. Department of Transport (1982) BD/14/82.

12. British Standards Institution (1978) *BS 5400: Steel, Concrete and Composite Bridges: Part 2: Specification for Loads.*

13. EMERSON, M. (1977) Temperature differences in bridges: basis of design requirements. TRRL Laboratory Report 765. Transport and Road Research Laboratory, Crowthorne.

14. CAPPS, M.W.R. (1968) The thermal behaviour of the Beachley Viaduct/Wye Bridge. RRL Report LR234, pp. 46. Road Research Laboratory, Crowthorne.

15. EMERSON, M. (1968) Bridge temperatures and movements in the British Isles. RRL Report LR 228, pp. 38. Road Research Laboratory, Crowthorne.

16. CAPPS, M.R.W. (1965) Temperature movements in the Medway Bridge. Interim report. Laboratory Note No LN/914/MWRC. Road Research Laboratory, Harmondsworth. (Unpublished).

17. EMERSON, M. (1966) Temperatures movements in the Hammersmith Flyover. Technical Note No. 33. Road Research Laboratory, Harmondsworth. (Unpublished.)

18. CAPPS, M.W.R. (1965) Thermal bending of the river span of the Medway Bridge. Technical Note No. 20. Road Research Laboratory, Harmondsworth. (Unpublished.)

19. EMERSON, M (1973) The calculation of the distribution of temperature in bridges. TRRL Report LR 561. Transport and Road Research Laboratory, Crowthorne.

20. ANON. (1986) Foyling the freeze. *New Civil Engineer*, 30 January, 23.

21. Reference 15, p. 8.

22. EMERSON, M. (1976) Bridge temperatures estimated from the shade temperature. TRRL Report LR 696. Transport and Road Research Laboratory, Crowthorne.

23. STEPHENSON, D.A. (1961) Effects of differential temperature on tall slender columns. *Concrete and Constructional Engineering*, **56**(5), 175–8: **56**(11), 401–3.

24. GARRETT, R.J. (1985) The distribution of temperature in bridges. *The Journal of the Hong Kong Institution of Engineers*, May, 35–8.

25. IHVE (1986) *Institution of Heating and Ventilation Engineers Guide, Book A, Design Data.*

26. Comité Euro-International du Béton (1984) *Design manual on structural effects of time-dependent behaviour of concrete* (Bulletin No. 142) George Publishing Company.

27. Comité Euro-International du Béton (1985) *Manual of Cracking and Deformations.* Bulletin 158E, Lausanne.

28. Federation International de la Precontrainte Recommendations (1984) *Practical Design of Reinforced and Prestressed Concrete Structures.* Thomas Telford.

29. British Standards Institution (1984) *BS 5400: Steel, Concrete and Composite Bridges: Part 4: Code of Practice for the Design of Concrete Bridges.*

30. NEVILLE, A.M., DILGER, W.H. and BROOKS, J.J. (1983) *Creep of Plain and Structural Concrete.* Construction Press, London and New York.

31. MATTOCK A.H. (1961) Precast–prestressed concrete bridge 5. Creep and shrinkage studies. *Journal of the Portland Cement Association Research and Development Laboratories*, May.

32. PILGRIM, D. and PRITCHARD, B.P. (1990) Docklands light railway and subsequent upgrading; design and construction of bridges and viaducts. *Proceedings of the Institution of Civil Engineers*, **88**, 619–38.

33. Institution of Geological Sciences: National Environmental Research Council (1976) *Atlas of Seismic Activity 1909–1968*. Seismological Bulletin No. 5.

34. DOLLAR, A.T.J., ABEDI, S.M.H., LILWALL, R.C. and WILLMORE, R.L. (1975) Earthquake risk in the UK. *Proceedings of the Institution of Civil Engineers*, **58**, 123–4.

35. ICE and SECED (1985) Earthquake engineering in Britain. *Proceedings of Conference of the Institution of Civil Engineers and the Society of Earthquake and Civil Engineering Dynamics*, University of East Anglia, April.

36. BAXTER, J.W., LEE D.J. and HUMPHRIES, E.F. (1972) Design of Western Avenue Extension (Westway). *Proceedings of the Institution of Civil Engineers*, **51**, 177–218. (Paper No. 7469.)

37. RICHMOND, B. (1980) Designing for temperature effects in concrete offshore oil-containing structures. Report UR17. UEG Construction Industry Research and Information Association, London.

38. LEE, D.J. (1967) Bending moments in beams of serial construction. *Proceedings of the Institution of Civil Engineers*, **38**, 621–37. (Paper No. 7023.)

TWO

Flexibility and articulation

2.1 General

Bridge structures must be designed to have sufficient flexibility to permit thermal and certain live load strains to occur without distress. This is recognized by BS 5400: Part 4 [1] which states

> The accommodation of movements and the method of articulation chosen … should be assessed … using engineering principles based on elastic theory…

Inherent restraints must be assessed and their effects minimized unless they induce a beneficial preload or are required to provide stability, as in the case of column bracing or plate stiffeners. Neglect of these strains may lead to expensive maintenance problems or failure by fatigue of steel or brittle fracture of concrete, due to biaxial or triaxial tension. Modern structural forms utilizing welded steel or prestressed concrete are not only frequently more complex and torsionally stiff, but are also more flexible and ductile than their older riveted mild steel or reinforced concrete counterparts. This flexibility may be internal, external or variable with the type and direction of movement.

A particular example of internal flexibility is the corrugated steel web plate which is stressed in shear by dead and live loads and provides buckling stability, but is not affected even by severe differential temperature strains.

An example of external flexibility is the system of vertical links commonly used at the ends of suspension bridge stiffening girders. These permit free longitudinal and plan rotational movements while providing vertical restraint.

Variable flexibility is not inherent, but may be introduced by mechanical devices such as hydraulic couplers between simply supported spans which permit free movement at slow rates of creep due, for example, to temperature changes, but are effectively rigid under dynamic loading. Hydraulic bumpers at railway bridge abutments act similarly [2].

Limit state design methods have been increasingly in general use since the publication of BS 5400 in 1984. However, these may not be appropriate for designs utilizing new materials which have high strength but low ductility. The possibility of sudden collapse of a structure composed of such materials can be removed by introducing temporary flexibility in the form of a 'fuse plug' member of a ductile material to provide stability at the collapse limit state.

2.2 Flexibility of the bridge structure

The contribution of individual components of a bridge structure to its overall flexibility are summarized in Fig. 2.1 and discussed in more detail in the next sections. It is necessary to consider the combined effect of these and the potential total resultant movements to determine the articulation required, particularly where a change of geometry will influence the load distribution.

The first possibility to be considered is whether a structure may be arranged without joints and bearings so that thermal changes, concrete shrinkage etc. can be accommodated by restraint of the structure and elastic deformation. In the past many brick and masonry viaducts were constructed in this way. The range of shade air temperature and effective bridge temperature in the northern parts of the UK is less than in the south (53°C to 57°C). Clearly, climatic conditions and form of construction can weigh very heavily in the engineering of short unjointed spans.

Concrete bridge spans up to 9 or 12 m can be built without bearings. With precautions for rotation, highway surfacing can be carried over the joints between beam and abutment for spans of up to 18 m.

Brick arch viaduct construction is another historical answer where the structure is capable of distortion without imposing excessive stresses in the structural elements or force on the restraining abutments. Probably the commonest modern engineering use of a fully restrained structural element is in long welded rail track.

At the other extreme, the advantages of building very long strings of spans with full continuity and freedom of movement poses the problem of designing bearings and expansion joints with a very large capability at the free end. The bearings and expansion joints for this type of structure have received considerable attention in recent years.

Fig. 2.1 (*Right*)   Relative flexibility of different bridge structures, (a) horizontally, transverse to bridge deck; (b) horizontally, longitudinally to bridge deck; (c) vertically.

2.3 Flexibility of the bridge components

2.3.1 SUPERSTRUCTURE

(a) Flexibility in the horizontal plane

The bridge deck is normally much stiffer in the horizontal plane than its support structure and so has little influence on the choice of the type of bearings or expansion joints. When the roadway is curved it is frequently advantageous to have the two carriageways on independent superstructures which can follow different vertical alignments. Twinning the decks in this way decreases the total horizontal stiffness by a factor of up to four. The longitudinal joint between the two must be designed to accommodate longitudinal shear movements between them due to thermal or other strains and for differential live load flexure.

Transverse movement joints can cause maintenance problems and poor riding characteristics so their number should be kept to a minimum compatible with the complexities of the joint and its seatings. Provision will also be required for drainage and services crossing the joints which can include bellows and thrust blocks at abutments if there is a significant change of direction in a pressurized pipe. Clearly such items must be identified at an early stage. With longer multispan continuous structures the large movement bearings required over many of the piers toward the free end can be difficult to accommodate and the end movement joint will be complex. An intermediate solution may be provided by linking together only groups of spans in lengths generally found to be an optimum of 350 to 400 m.

In the Mancunian Way elevated structure [3], which has a compositely curved horizontal alignment, an investigation was made into the forces acting on the bearings on the columns during movements of the structure. A series of computer analyses were made to establish the effect of flexibility in the substructure, omission of restraints at intermediate columns, variations in friction coefficient, clearances in the bearings, and optimum alignment of the bearings. A plane frame program treated the structure as polygonal and replaced the tapering columns by spring members with stiffness appropriate to their height. It was found

	Superstructure	Connection between superstructure and substructure	Substructure	Connection between substructure and foundation	Foundation
Stiffest	Deep edge beams	Encastre	Wall pier	Encastre	Spread footing on rock
	Multicell	Rigid bearings	Raking leg frame pier		Caisson
	Multibeam		Portal frame pier	Pinned bearings	Spread footing on soil
	Spine box	Sliding bearings with slide restraints			Raking piles
	Flat slab		Narrow pier		Vertical bored cylinders
					Vertical piles
Most flexible	Tin decks	Free sliding bearings	Multicolumn pier	Sliding bearings	Vertical slip plane footing
Stiffest	Straight continuous spans	Encastre	Raking leg frame	Encastre	Longitudinal shear walls
		Pin bearings	Arch		Caisson
	Gently curved continuous spans	Fixed rocker bearings			Spread footing
			Short column		
	Simply supported spans tied in groups	Elastomeric bearings	Portal frame		Raking piles
			Encastre slender column		Vertical bored cylinders
	Sharply curved continuous spans	Free sliding bearings	Twin shaft column		Vertical piles
Most flexible	Simply supported spans	link bearings	Pinned column	Pinned	Friction slip plane footing
Stiffest	Deep continuous girder spans		Wall piers		Rock
			Vertical columns		Stiff soils
	Cantilevered girders with suspended spans		Raking column frames		Compactable soils
Most flexible	Simply supported spans		Two pinned portal frames		Ground liable to mining or other settlement

convenient to consider the resulting forces under two headings: firstly, the spring forces generated to constrain the structure to move through the expansion joint at the free abutment if bearing friction is ignored; and secondly, the forces resulting from bearing friction.

The spring forces due to thermal, creep and shrinkage movements would be zero if the structure were free to expand in any direction on frictionless bearings. In practice, lateral restraints are required to carry wind and centrifugal forces, even if these are not necessary to control the alignment of the structure at the expansion joint. Initially the structure moves freely to take up lateral clearances in the bearings. Subsequent movement produces an equilibrium set of restraint forces in the bearings. As can be seen from the typical restraint force curve in Fig. 2.2, the sign of the forces changes several times. Thus, the bearing clearances on opposite faces must be closed before a balanced set of forces can develop, and this must be borne in mind when combining

restraint forces with the unidirectional wind and centrifugal forces. The force developed at the bearing guide is related to the angle between it and the direction of free movement. This will result in a friction force which is additional to that existing due to the vertical load of the structure. Creep and shrinkage movements are irreversible, so they can dominate the pattern in which lateral bearing clearances close (they should have been set equal at the time of construction) and the resulting restraint force system.

The Mancunian Way investigation proved that an increase in the bearing clearances or a reduction in the substructure stiffnesses causes a reduction in the restraint forces. It was also shown that, in theory, the forces can be minimized by orientating the bearings on the column caps to lines which are not necessarily tangential to the structure. However, for a multicurved structure, the optimum line is not normally worth the trouble of setting out.

In a long straight structure, the thrust which accumulates from the forces due to bearing friction causes no lateral load on the bearings. In a curving structure, however, it was shown that significant transverse forces arise which depend on the angular deviation of the structure (see Fig. 2.3). The friction forces, which increase towards the fixed end, do not form a balanced set and are not influenced, apart from local redistribution, by variations in bearing clearance or substructure stiffness. The build up of friction forces varies in a random pattern dependent on the slip/stick situation at each bearing.

The investigation assumed that the bearings would be installed with the sliding surfaces horizontal to avoid any wedging action during translation.

(b)   Flexibility in the vertical plane

For normal vertical alignments in which grades do not exceed about 1 in 20, the load/deflection characteristics of the superstructure in the vertical plane are such that force components generated by horizontal movements due to thermal or other strains in the superstructure can be ignored.

Halved joints incorporating bearing seatings or linkages are commonly used at the ends of suspended spans to allow flexural and other

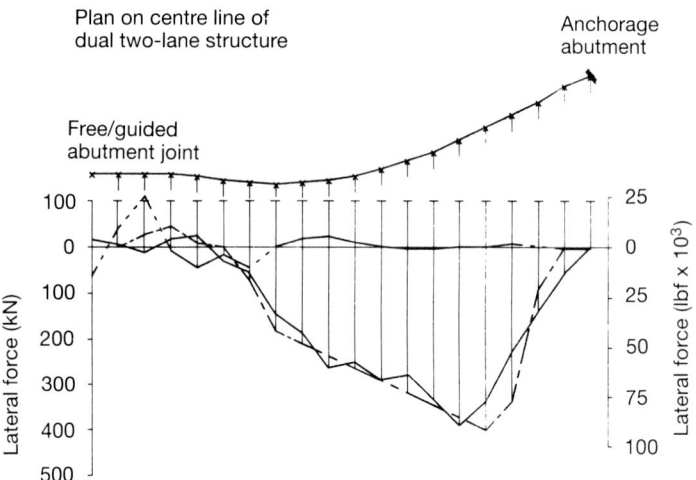

Fig. 2.2 Distribution of lateral forces on columns due to movement of deck space. ——— : friction force, flexibility of columns considered; – – – : friction force, flexibility of columns not considered; – – – – : expansion force.

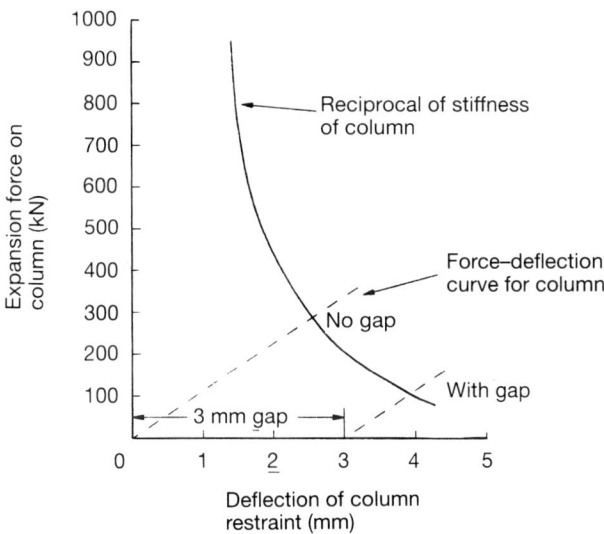

Fig. 2.3   Effect of a gap on the expansion force of a typical column of the Mancunian Way.

movements, but are not recommended unless special access provisions are made for bearing maintenance and repair. This is usually easier to arrange in a steel than a concrete superstructure. However, the problems which can arise have been illustrated by the failure of the bearing linkages on the Mianus Bridge in the USA, reportedly due to forces arising from the skew of the span which had not been allowed for in the design. Corrosion between link or pin plates can obviously result in loss of section, but interface bursting pressures of at least 20 N/mm^2 can also be generated potentially, causing failure of pin end plates or the seizing up of the linkage.

2.3.2   CONNECTION OF SUPERSTRUCTURE TO SUBSTRUCTURE

Either restrained or movement bearings are most commonly adopted at this interface, but accessibility for inspection and maintenance can be a problem unless suitable provision is made at the design stage. When the deck width exceeds 5.5 m, allowance is required for transverse movements at the bearings unless clearances within them exceed 2 mm. It is then advisable to locate the superstructure by one of the bearings only at each pier or abutment, leaving the others free to move transversely.

In the past it was usual to allow for rotational movements by rocker plate surfaces or pins. However, elastomeric bearing pads which require less maintenance are now commonly adopted for short and intermediate span structures. These may be unreinforced where only small rotations are applied or reinforced by steel plates if the rotations are greater or combined with significant shear movement. For intermediate spans the elastomeric pad is fully contained in a pot to cater for rotational flexibility independently from sliding surfaces allowing shear movements. It should be noted that the stiffness of an elastomeric bearing is not only related to the composition of the bearing but also to the rate of load application. Spherical bearings also provide complete rotational flexibility but have limited capacity to resist horizontal forces.

Tie down linkages between the superstructure and substructure provide flexibility for both the rotational and longitudinal movement arising in cantilever and cable-supported structures. They can also incorporate provision for transverse rotation when free torsional flexure of the structure is required, but tend to become complex and bulky, and are best avoided if possible. Fatigue-prone high tensile materials are unsuitable for use in linkages unless prestressed to overcome the effects of stress fluctuations.

Frame action with the superstructure fixed to the substructure can produce economy, but must allow for both the reversible (temperature, transient load, construction) and the irreversible (creep, shrinkage, settlement) effects.

2.3.3   SUBSTRUCTURE

(a)   Short columns

A short column may be defined as one which will not fail by local buckling. In BS 5400: Part 4 [1] a concrete column is considered as short if the ratio l_e/h is less than 12. Its Euler critical stress can be written as

$$P_{cr} = \frac{\pi^2 EI}{l_e^2} = \frac{\pi^2 Eh^3}{12l_e^2}$$

or $$\sigma_{cr} = \frac{\pi^2 E}{2}\left(\frac{h}{l_e}\right)^2$$

For $l_e/h = 12$ this will be in the range 140 to 200 N/mm^2 depending on the grade of concrete used. The rotational restraint assumed in the derivation of the effective height l_e is quoted in BS 5400 (Table 2.1) from which it can be seen that a maximum ratio of $l_e/h = 12/0.7 = 17.1$ would apply with full fixity at both the superstructure and the foundation. In practice this can never be achieved because the deck has a degree of flexibility. For columns built into the deck the correct flexibility ratios would develop from a frame analysis, but the foot of the column should similarly receive realistic treatment.

The distribution of traction forces along a viaduct with flexible columns and bearings can be simply derived by means of the graphs in Fig. 2.4. These are based on the assumption that the viaduct is long enough for effects at either end to be independent. The deck members are mounted on flexible bearings at one end of each span and fixed bearings at the other.

If we apply a traction force P to the span connected to some column X near the centre of the viaduct, then the load carried by this column is βP. The load carried by columns adjacent to X falls off in a geometric progression of the form

$$\frac{\beta P}{U}, \frac{\beta P}{U^2} \cdots \frac{\beta P}{U^n}$$

where

$$U = \left(1 + \frac{R}{2}\right) \pm \sqrt{R + \left(\frac{R}{2}\right)^2}$$

The value of β is plotted against R in Fig. 2.4, where $R = Z/S$, the ratio of the column stiffness (Z) to the sum of the shear stiffness (S) of all the bearings under the loaded span.

If an abutment of a long viaduct is rigid, and the end span is supported on it on flexible bearings, the load carried by the end column when a traction force P is applied to the end span is γP, and the force transmitted to the abutment is θP.

If the abutment is flexible (the same stiffness as the columns) and a traction force P is applied to it, the load carried by it is αP, and the load carried by the end columns is $\alpha P/U$.

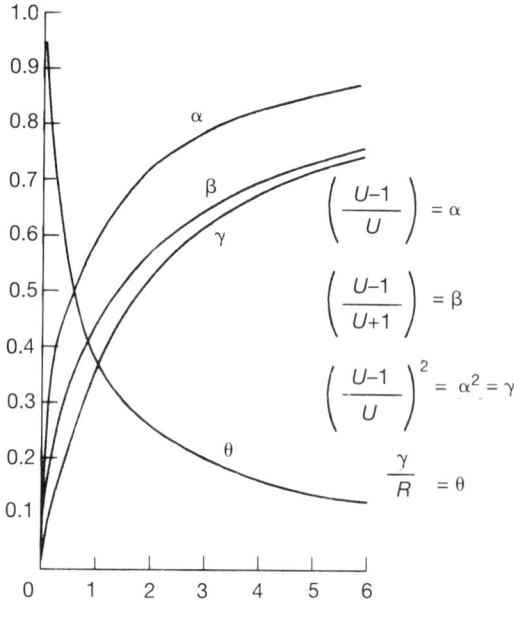

$$\left(\frac{U-1}{U}\right) = \alpha$$

$$\left(\frac{U-1}{U+1}\right) = \beta$$

$$\left(\frac{U-1}{U}\right)^2 = \alpha^2 = \gamma$$

$$\frac{\gamma}{R} = \theta$$

Ration of stiffness R=Z/S

Fig. 2.4   Relation between proportion of load carried by column and stiffness ratio.

Table 2.1 Effective height, l_e, for columns (extract from BS 5400: Part 4: 1984)

Case	Idealized column and buckling mode	Location	Restraints Position	Rotation	Effective height, l_e
1		Top	Full	Full[1]	$0.70l_0$
		Bottom	Full	Full[1]	
2		Top	Full	None	$0.85l_0$
		Bottom	Full	Full[1]	
3		Top	Full	None	$1.0l_0$
		Bottom	Full	None	
4	Elastomeric bearing	Top	None[1]	None[1]	$1.3l_0$
		Bottom	Full	Full[1]	

Case	Idealized column and buckling mode	Location	Restraints Position	Rotation	Effective height, l_e
5		Top	None	None	$1.4l_0$
		Bottom	Full	Full[1]	
6		Top	None	Full[1]	$1.5l_0$
		Bottom	Full	Full[1]	
7	or	Top	None	None	$2.3l_0$
		Bottom	Full	Full[1]	

[1]Assumption made in BS 5400: Part 4.

Zederbaum [4] has also considered the elastic distribution of forces and displacements. This approach is essential to a thorough assessment of structures incorporating elastic bearings.

Leonhardt and Andrä [5] have shown that it is possible greatly to relieve the fixed end moment applied to spread footings by taking into account the modulus of reaction of the soil as a spring loading. They suggest that with high modulus of reaction and a narrow foundation the assumption of a pin joint at the base is more correct than the assumption of fixity. In most instances, some degree of fixity is present and provided that the modulus of reaction of the soil is available there is no difficulty in analysis. If, however, soil characteristics are not available it is misleading to take the limiting encastre case and assume this is on the conservative side.

If the column is of concrete with a spread footing on rock or a large pile cap, it is reasonable to assume a supporting medium of similar elasticity to the material of the column. It is then quite simple to take into account this assumption of elasticity of the support.

Consider that the base of the column is built into a semi-infinite supporting material of the same type as the column (Fig. 2.5). When fixed-ended, the deflection at the top $pl_o^3 / 3EI = 12pl_o^3 / 3Eh^3$. If we put $l_o/h = \gamma$, then deflection $\delta_E = 4p\gamma^3 / E$. There is an additional deflection caused by the elasticity of the supporting medium. Roark [6] cites O'Donnell [7] to give $\Delta\Theta = \dfrac{16.67\, pl_o}{\pi\, Eh_1^2} + \dfrac{(1-v)p}{Eh_1}$ where $h_1 = h + 1.5r$. In this example, the root fillet radius $r = 0$. This additional deflection is $l_o\Delta\Theta$ at the top. Hence, deflection

$$\delta_F = \frac{16.67\, pl_o^2}{\pi\, Eh^2} + \frac{(1-v)pl_o}{Eh} = \left(\frac{16.67\,\gamma^2}{\pi} + (1-v)\gamma\right)\frac{p}{E}$$

The increased deflection at the top is thus in non-dimensional terms:

$$1 + \frac{\delta_F}{\delta_E}$$

where

$$\frac{\delta_F}{\delta_E} = \frac{(16.67\,\gamma/\pi) + (1-v)}{4\gamma^2}$$

For concrete. Poisson's ratio v is approximately 0.2. Hence,

$$\frac{\delta_F}{\delta_E} = \frac{5.3\,\gamma + 0.8}{4\gamma^2}$$

This relationship is illustrated in Fig. 2.5. It will be seen that at the limit for a short column pinned at the top the deflection can be 26% greater (15% with an elastomeric bearing at the top) than the classical bending deflection assuming full base fixity. This in itself reduces the required travel of bearings and the forces generated. Pile caps and pile foundations would generally be more flexible than the assumption of an infinite elastic medium, and this could be assessed separately by investigating the frame action of the pile group. It will be clear, however, that the actual rotation generated at the top of a column is greater than when base fixity is assumed.

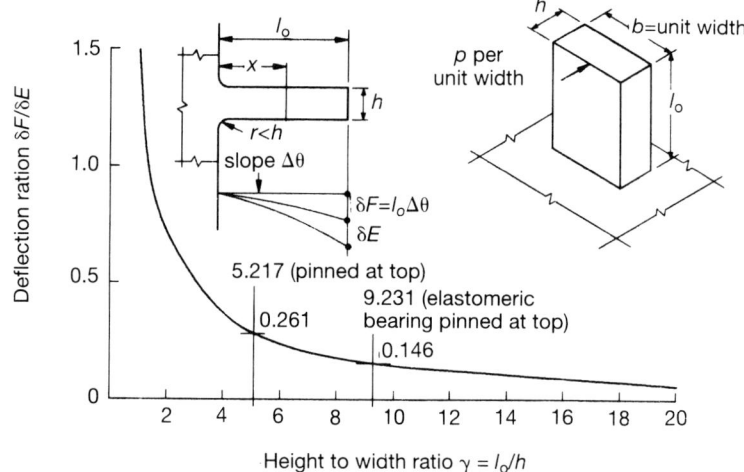

Fig. 2.5   Relation between deflection ratio and height/width ratio.

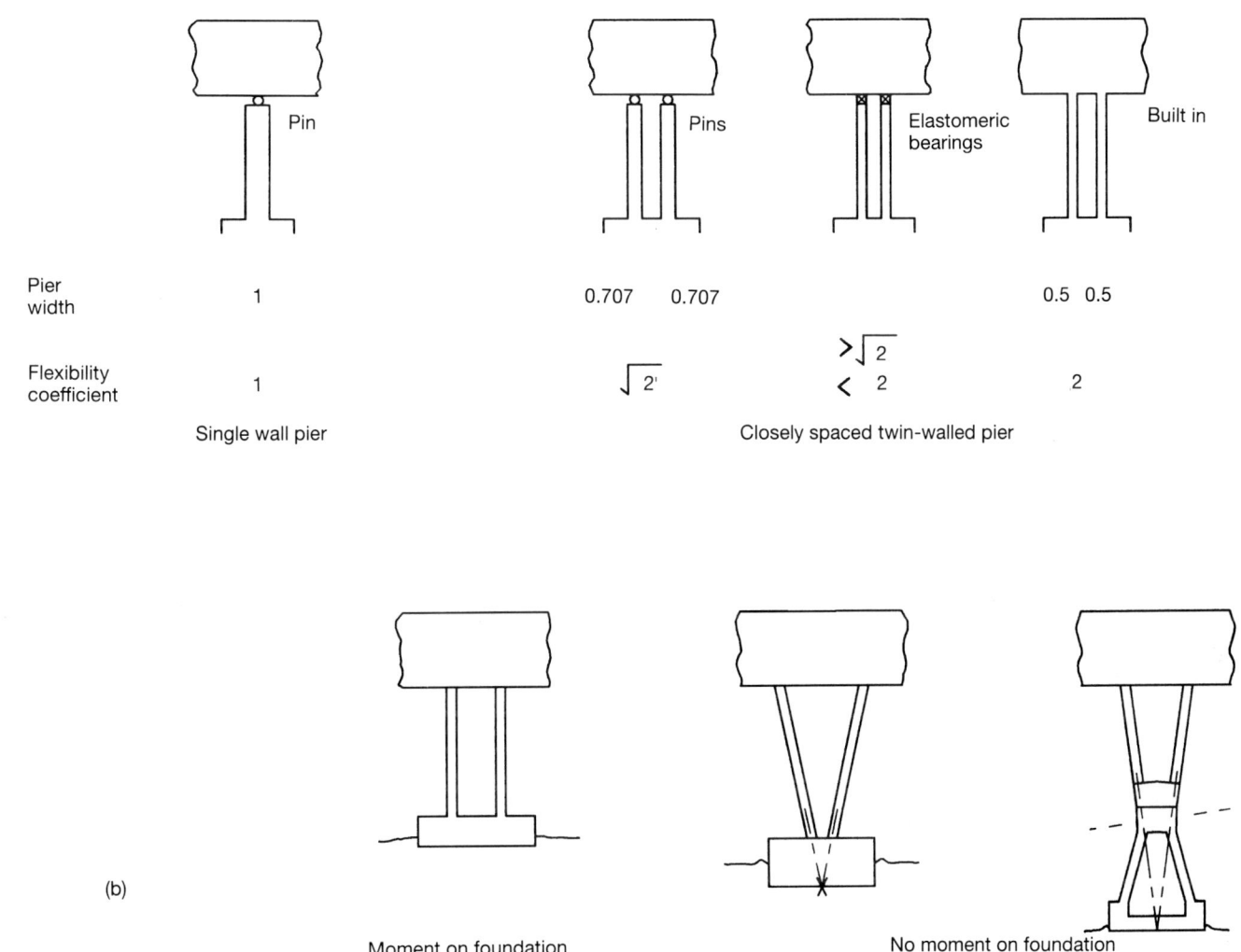

(a)

Single wall pier   Closely spaced twin-walled pier

(b)   Moment on foundation   No moment on foundation

Widely spaced twin walled pier

Fig. 2.6 Flexibility of twin-walled pier, (a) negligible interaction with deck; (b) frame action.

For the precise assessment of the behaviour of short columns, the elastic foundations can be properly dealt with by the insertion of spring supports for tilt, vertical and lateral translation such as given in standard texts ([8] p. 118).

Where forces are not reversible, the expressions for deflection after creep have also been given ([8] p. 877).

(b) Twin-walled piers

A slim, vertical pier poses the problem of flexibility for taking up movements whilst having sufficient elastic stability. However, increased pier flexibility can also be obtained by the adoption of twin walls. Provided their separation is small they are not significantly influenced by deck rotations or overall (portal) frame action. The range of flexibility which can be achieved in this way is summarized in Fig. 2.6(a).

If the spacing of the walls is increased (Fig. 2.6(b)) frame action develops and the limiting values in Table 2.3 are reduced due to the resulting changes in axial loads. These actions can be significantly modified by the introduction of rubber bearings on top of the walls. Increased moments are also applied to the foundation.

A particular application of this arrangement is in balanced cantilever concrete construction. Temporary bracing between the walls can be used to provide stability during cantilevering and is removed when the midspan deck connection is made.

If the increased moment on the foundation is unacceptable it can be eliminated by sloping the walls so that their centre lines intersect at foundation level. Muller [9] has shown how an optimum relationship of these factors can be achieved by the geometrical layout, with particular reference to the use of such twin inclined piers in the bridge over the Seine at Choisy-le-Roi. Features of this design include:

1. neoprene bearings on top of the walls;
2. end restraint over part of the height of the walls;
3. wall slopes 6.5%.

(c) Slender columns

In BS 5400: Part 4, a slender concrete column is defined as having

$$12 < \frac{l_e}{h} \leqslant 40$$

(or $\leqslant 30$ if there is no positional restraint at one end) where l_e is the effective height in the plane of buckling under consideration and h is the depth of the cross-section in the plane of buckling under consideration. In the present context it is of interest to consider how the effective length of a column is affected by spring stiffness at the bearing and in the foundation. These parameters plotted to logarithmic scales are shown in Fig. 2.7 in which the values shown in Table 2.2 are readily recognizable.

The values of effective length given in Table 11 of BS 5400 are seen as $K_F/K > 4$ with $K_s l^2/K$ large in cases 1 and 2 (l_e tending to $0.7\, l_o$) and $\frac{K_F}{K} > 8$ with $\frac{K_s l^2}{K}$ small in case 7 (l_e tending to $2.3 l_o$). So for a column of uniform cross-section the appropriate effective length can be obtained from this figure for any degree of top and bottom restraint.

A pier having sufficient flexibility to accommodate movements of the structure will require a slenderness ratio beyond the limit of 40 set by BS 5400: Part 4 [1]. In the Tasman Bridge, Hobart (Figs 2.8 and 2.9) opened in 1965 the columns of the approach spans are from 12 to

Table 2.2 Parameters plotted in Fig. 2.7 for cases given in BS 5400: Part 4, Table 11

Case	$\frac{l_e}{l}$	$\frac{K_s l^2}{K}$	$\frac{K_f}{K}$
1	0.7	∞	∞
2	0.85	∞	4
3	1.0	π^2	0
4	1.3	~ 4.5	8
5	1.4	~ 3.8	8
6	1.5	~ 3	8
7	2.3	0	8
—	π	1.0	0

Table 2.3 Relative flexibility of short single- and twin-walled piers

| Case | Maximum stress (for $b_1=1$) | Maximum deflection (for $b_1=1$) | Equivalent pier thickness for the same maximum stress | Examples for $h=10b$ and $b_1=1$ | | | | |
| | | | | H/P | | | | |
				0.05	0.1	0.2	100	∞
1. Single-wall rectangular section pier	$\dfrac{P}{K_1}+\dfrac{6M}{K_1^2}$	$\delta_1=\dfrac{4Mh^2}{EK_1^3}$	$K_1=1$	$K_1=1$ $\delta_1=1$	$K_1=1$ $\delta_1=1$	$K_1=1$ $\delta_1=1$	$K_1=1$ $\delta_1=1$	$K_1=1$ $\delta_1=1$
2. Twin-wall pier pinned at top	$\dfrac{P}{2K_2}+\dfrac{3M}{K_2^2}$	$\delta_2=\dfrac{2Mh^2}{EK_2^3}$	$K_2=\dfrac{P+\sqrt{P^2+3M(P+6M)}}{4(P+6M)}$	$K_2=$ 0.6757 $\delta_2=$ 1.6208 δ_1	$K_2=$ 0.6906 $\delta_2=$ 1.5181 δ_1	$K_2=$ 0.6987 $\delta_2=$ 1.4646 δ_1	$K_2=$ 0.7069 $\delta_2=$ 1.4155 δ_1	$K_2=$ $1/\sqrt{2}$ $\delta_2=$ $\sqrt{2}\,\delta_1$
3. Twin-wall pier built in at top	$\dfrac{P}{2K_3}+\dfrac{3M}{2K_3^2}$	$\delta_3=\dfrac{Mh^2}{EK_3^3}$	$K_2=\dfrac{P+\sqrt{P^2+1.5(P+6M)}}{4(P+6M)}$	$K_3=$ 0.4966 $\delta_3=$ 2.0414 δ_1	$K_3=$ 0.4990 $\delta_3=$ 2.0121 δ_1	$K_3=$ 0.4997 $\delta_3=$ 2.0036 δ_1	$K_3=$ 0.5000 $\delta_3=$ 2.0000 δ_1	$K_3=$ 1/2 $\delta_3=$ 2δ_1

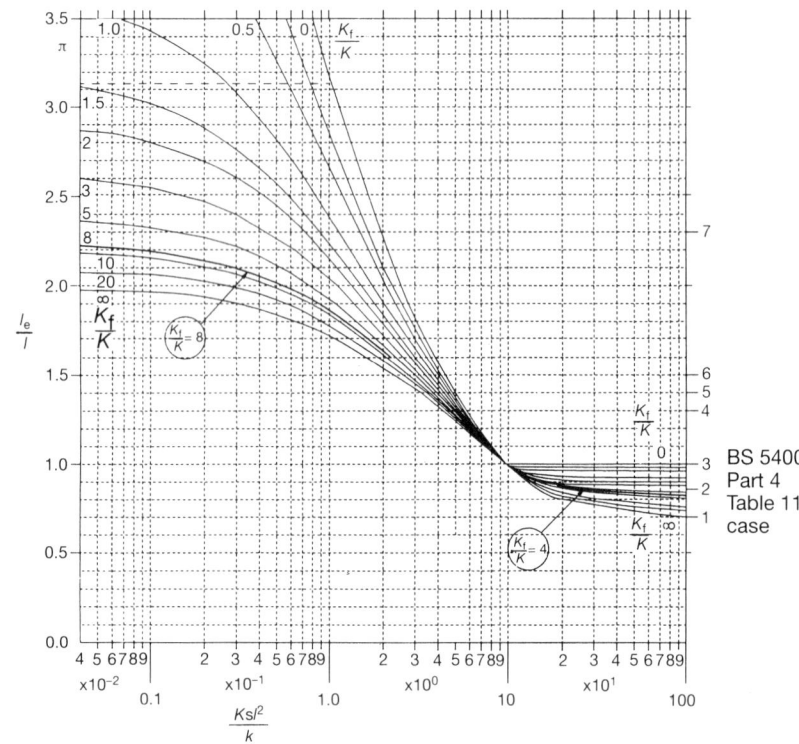

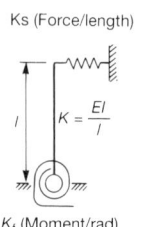

Fig. 2.7   Effects of spring stiffness at ends of slender columns.

38 m high springing from a pile cap at water level and with a steel rocker bearing on top absorbing rotation changes in the continuous prestressed concrete deck but providing positional fixity. The columns are high enough for the assumption of full fixity at the base to be perfectly valid. The bending deflection of the concrete columns provides the movement capability. The slenderness ratio is up to 50, well outside the BS 5400 provisions for columns.

However, with this articulation, the column is determinate and can be designed for the actual moments obtained. The design may be based on either the ultimate strength of the section or the yield stress of the material, with a suitable load factor. The analysis was based on normal elastic theory, resulting in the design curves shown in Fig. 2.10 which are self-explanatory.

The analysis for the bridge column design also took account of lack of straightness in construction and accidental eccentricity of the vertical prestress which was provided to ensure that the concrete columns exhibit proper elastic behaviour.

In many cases it will be necessary to assess effective lengths using a more rigorous buckling analysis. This analysis should make an allowance for the effects of the change of the geometric form of the structure (e.g. the effect of axial loads on deflections), changes in stiffness of materials under load (e.g. cracking of concrete), lack of straightness in construction and eccentricity of bearings. It should also take account of support provided at the base and at the intermediate points.

Methods which obviate the need to calculate effective lengths for concrete cantilevers are given below. However, for other concrete structural forms the buckling analysis should use elastic cracked section properties where vertical loads are not high enough and members are too slender to prevent cracking.

The basic method follows BS 5400; Part 4 Clause 5.5. The moment at the base of the column is

$$M = M_i + Ne_i + Ne_{add}$$

where M_i is the applied moment from horizontal loading; N is the axial load; e_i is initial actual eccentricity for the unloaded pier.

Fig. 2.8 Tasman Bridge, Hobart, Tasmania. View of elastically flexible column.

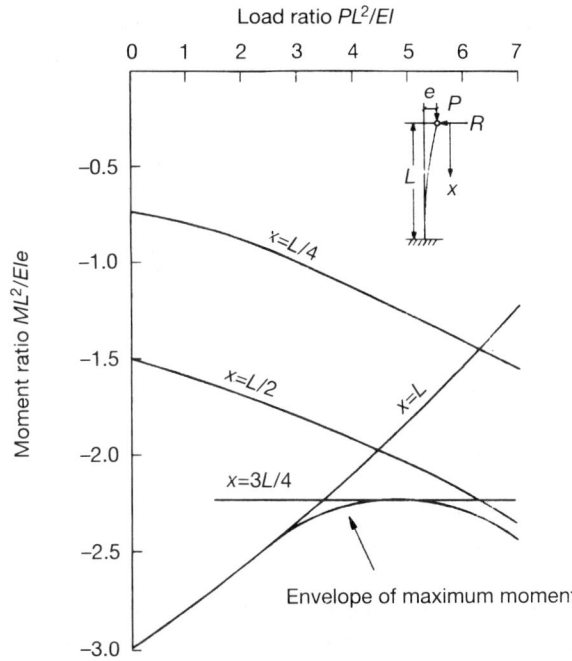

Fig. 2.10   Design curves for deflected column under normal load.

$$e_{add} = \frac{l_e^2}{1750h}\left(1 - 0.0035\frac{l_e}{h}\right)$$

= displacement of reinforced concrete section at ultimate limit state.

The additional eccentricity (e_{add}) can be very large so some saving is possible in cases with high axial load. The methods of Clause 3.8.3.1 of BS 8110: Part 1 may then be adopted.

An alternative method for slender cantilevers is to replace e_{add} by $(e_{a1} + e_{a2})$ where

$e_{a1} = e_{add}$ as above but $l_e = 2l_o$;
e_{a2} = displacement due to base rotation = Ml_o/K_f
where K_f = base stiffness (moment/radian).

$$M = M_i + N(e_{a1} + e_{a2})$$

hence

$$M = (M_i + Ne_i + Ne_{a1})/\left(1 - \frac{Nl_o}{K_f}\right)$$

Serviceability limit state conditions should be checked if the l_e/h limits of BS 5400 are not satisfied. This will require the checking of elastic stresses and crack-widths. The design moment should include the additional moments due to e_{a2} and e_{a3} where e_{a3} is the elastic deflection of the column using cracked section properties.

(d)   Multiple supports

Individual support columns under each bearing in the cross-section give a light appearance to the structure. In addition they give increased flexibility which will reduce the transverse loads on bearings supporting curved continuous superstructures. However, they are not appropriate for the submerged portion of a river pier where they would disturb the river flow, increasing bed scour, and possibly trap floating debris.

Tubular steel columns are outside the scope of BS 5400, but these members can be designed in accordance with the Det Norske Veritas (DNV) design rules for offshore structures (in particular [10]). Some modifications will be necessary to convert it to a limit state format, with appropriate levels of reliability for bridge design with a 120-year life.

Concrete-filled tubular steel members should be designed as reinforced concrete members with external reinforcement. Shear connection is needed between the infill concrete and the steel tube.

Fig. 2.9   Tasman Bridge.

2.3.4   CONNECTION OF SUBSTRUCTURE TO FOUNDATION

The connection of piers to the foundation is most commonly rigid to simplify construction, and also to allow submergence in the case of river piers. However, a pinned connection may be required to avoid transmitting a moment on to the foundation slab, deep flexible pile or cylinder foundation at, for example the foot of a portal frame leg. In addition it may be appropriate to introduce a controlled friction plane (e.g. by a sand layer or felt on smooth blinding) where horizontal seismic or subsidence movements can be allowed for.

Bearings allowing rotation and longitudinal movement were used under the portal frames supporting one section of the Western Avenue Extension elevated road (Fig. 2.11). These were required by the rigid connection of the portals to the superstructure deck in order to achieve a minimum construction depth for the composite deck and portal crosshead.

2.3.5   FOUNDATION

Complete rigidity in the foundation is impossible to achieve and is only approached by a connection of the substructure to rock on a wide base. The degree of flexibility increases with slab bases and caissons on soils, but not to an extent which will influence the design of bearings and movement joints except when ship impact forces or seismic effects have to be considered. Flexibility may, however, come into play when deep cylinders or piles pass through a substantial depth of soft material. This may have sufficient stiffness to restrain them against buckling instability while permitting a degree of flexural movement and rotation. A logical conclusion might then be to extend the cylinder out of the ground as a pier and eliminate any bearing or joint by making it integral with the superstructure. However, it is rare to find a situation where the conflicting requirements of movement, strength and stability can all be met. Caution is also required in depending upon soil characteristics which could change or be irreversible as a result of ground water movements under sustained load.

2.4   Articulation of the bridge structure

From an articulation point of view the ideal is achieved by eliminating mechanical joints and bearings and accommodating movements and rotations by internal flexibility as seen, for example, in old brick viaducts. However, for a modern bridge this is not normally a practicable possibility. A number of different methods of providing articulation are shown in Table 2.4 together with notes summarizing their advantages and disadvantages. It is seen that the articulation may be provided by joints or member flexibility in a variety of ways. How then is the most appropriate arrangement to be identified?

The first question to be answered is whether differential settlements are likely. If this is not the case then a continuous superstructure would be expected to be economic and this will often be the case for flyover structures. Limited differential settlement due to, say, mining subsidence can be tolerated by providing jacking pockets to allow periodic adjustment. In river crossings, however, the founding depths and materials may differ between individual piers and at the abutments and complete continuity of the deck may not be appropriate. For longer spans, cable-stayed and suspension bridges are found to have different requirements. In the former, articulation at the ends and continuity at the towers are commonly preferred because of the balanced horizontal components of the cable loads applied to the deck, but for the latter, joints at the towers are preferred to limit movements on the shorter hangers at midspan and near the ends of the bridge.

Next to be considered is whether it is more economic to adopt a single large anchorage for longitudinal forces at one pier or abutment or to distribute the load between the piers and abutments by balancing their flexibilities, and possibly also by using elastomeric pad bearings. For relatively low structures it may additionally be economic not to restrain the superstructure at every pier to carry transverse overturning from wind or centrifugal forces, but to take advantage of the horizontal strength of the deck to span between selected piers.

Fig. 2.11   Westway, Section 6, London.

For railway bridges, continuity requiring larger movement joints may require special joints in the track over them and for that reason may be unacceptable.

When provision for the horizontal effects of seismic action or mining settlement is required, it may be provided by controlled sliding friction planes. The movement at articulation joints then requires special consideration.

Table 2.4   Alternative arrangements of bridge articulation

Arrangement[1]	Advantages	Disadvantages
Simply supported span	Small movements of joints; accommodates differential settlements	Large number of joints, width of bearing seating on piers
Simply supported and drop-in span	Small movements of joints; accommodates differential settlements	Difficult access for maintenance of halved joints
Continuous spans fixed at one end.	Minimum number of joints	Pier settlement induces deck bending moments; large movement of joints; large overturning moment on one abutment

Table 2.4 (contd)

Arrangement[1]	Advantages	Disadvantages	Arrangement[1]	Advantages	Disadvantages
Multispan rigid frame fixed at one end	Minimum number of joints; ease of inspection and maintenance of bearing; minimizes overturning moments on foundations	Pier settlement induces deck bending moments; large movement joint; large overturning moment on one abutment	Rigid frame with fixed central span	Suited to balanced cantilever construction	Possible pier instability during erection; possible uplift at abutments
Continuous spans on rocker columns	Minimum number of joints; minimizes moments on foundations	Pier settlement induces deck bending moments; piers require support during construction; large movement joint; large overturning moment on one abutment; risk of progressive collapse	Continuous spans on elastomeric bearings	Longitudinal forces shared between supports; low maintenance of elastomeric bearings	Limited by movement possible on elastomeric bearing; requires symmetry in support structure; pier settlement induces deck bending moments; requires provision for bearing replacement
Continuous spans fixed at central support column	Minimum number of joints	Anchorage foundation at centre of bridge	Wichert truss	Continuity of structures even if piers subside	Cost of pin connections in truss over the piers

[1]Bearings: ×, rocker/pot fixed; ∞, guided roller/sliding pot; ●, pin; ■, elastomeric. Joints: R, rotation only; M, movement and rotation.

2.5 Traditional arch construction

Reference has already been made to the many thousands of traditional arches giving service in old established countries like the UK. With increasing frequency or intensity of highway traffic it is necessary to review the durability and load carrying capacity of old stone and brick arches. As might be expected with a form of historical construction thousands of years old there is a great deal of empirical knowledge but little rigorous scientific observation and theory. It has been necessary to address the problem in recent times.

A basic theory of arches does not deal with the problem of arch bridge construction. For example a medieval arch bridge comprises:

1. the arch ring itself;
2. the springing, abutments and intermediate pier, if any;
3. the spandrel walls and filling;
4. the foundation.

Multispan railway viaducts and the like also introduce the behaviour of the intermediate piers, some very tall. Old bridges have usually been modified and/or widened during their life with a consequent further interaction of structural behaviour.

An arch bridge is therefore a complex three-dimensional structure. The simple arch does not consider the stiffness of spandrels. For a narrow bridge the two spandrels could carry much of the load as an overhand cantilever construction or through girder. Several types of structural action are called upon either concurrently or independently.

The filling above the arch rings may or may not be composite. Spread, settlement and tilt of piers and abutments is very significant and can well master the arch action itself.

Every encouragement is given for the engineer to seriously consider traditional arch construction in appropriate cases. Recourse to the developing modern literature and research in this field is most interesting and helpful [11,12].

Whilst most abutments and piers are of concrete it is relevant to note that brickwork and masonry may be considered either as facing or fully structural.

2.6 Integral bridges

The term 'integral bridges' has come into general parlance and it is necessary in this volume to consider the issues that are raised. Briefly it is important to recognize what the term is intended to embrace. Similarly it is important to accept the limitations that the concept imposes.

It is a basic premise in this book to view continuity as a helpful concept in creating bridges of good serviceability and adequate durability. Considerable detail has been attempted to allow the designer to evaluate the physical movements and constraints that are developed in any particular bridge scheme.

What must be emphasized is that the physics of the materials such as steel and concrete dominates our understanding of how to design bridge structures. Thus long span suspension bridges and long multi-span viaducts can be of continuous construction, but it has to be recognized that temperature movements from such structures cannot be ignored. Temperature movements of the order of over a metre cannot be forgotten but must be accommodated by expansion joints.

This is similar to relating the change of strain to a stress condition which in long lengths is greater, sometimes by many times the permissible stress of the material.

The other major aspect is the form of the bridge. Reference has been made to tall, slender columns, frames and arches which rely on the change of strain due to temperature being absorbed in the flexural distortion of the structure. It is obviously helpful if unnecessary joints are eliminated from frame and arch construction through the evaluation of the appropriate stress condition under change of shape.

The use of integral bridges is therefore concerned with the design of bridges which do not require expansion joints. They are bridges which do not present excessive strut/tie forces and have shapes that allow the dispersion of strain changes.

In practical terms therefore the integral bridge concept has been tried out in the USA and Canada as particularly applicable to freeway overpasses of one, two or four spans. The bridge spans themselves present no difficulty, but if they are fully connected to the abutments then clearly the behaviour is dependent on the form of the abutments.

If the abutment is a type of bank seat attached to the bridge then a whole body motion of the bank seat is implied. The soil should be able to absorb this movement and correct run on slabs designed. If a wall type abutment is attached the wall flexure is brought into play.

It is then necessary for there to be sufficient space behind the abutment for the subsoil under the approach to take up the strain. This in turn requires the use of run-on slabs designed to span across that part of the fill behind the abutment to prevent traffic compacting soil that is constantly disturbed to some extent by the movement of the abutment.

A further detail is that required between the rear end of the run on slab and the highway construction. A similar transition is required on railway bridges. With ballasted track this is similar to highway construction. For bedded track a considered design of transition is required.

It could be argued that the integral concept moves the problem of expansion joints from the bridge to the approaches. North American opinion suggests, for example, that it is cheaper and easier to repair damage to the junction between run-on slab and the highway itself than it is to repair damage to the expansion joints on the bridge deck. Where de-icing salts are used, leakage of joints promotes the spread of damaging chloride-laden liquid. Chloride impregnation damage to concrete elements of the bridge is obviously more troublesome than restoring highway construction at grade. It is also argued that actual movements that take place daily are much less than theoretically predicted or are cushioned and it is only the seasonal or extreme variations that may give rise to occasional limit state conditions.

A designer should therefore refer to the relevant evaluation of most benefit to assist the decision making process. This is, of course, in addition to the normal structural evaluation discussed elsewhere in this book.

Such matters require consideration and include *inter alia*:

1. database of climatic effects in the area;
2. soil/structure interaction evaluation;
3. intensity of traffic and whether urban or rural location;
4. quality and budget of maintenance authority;
5. consequences of low performance;
6. accessibility for repairs and ease of lane closures etc.

References

1. British Standards Institution (1984) *BS 5400: Steel, Concrete and Composite Bridges: Part 4: Code of Practice for the Design of Concrete Bridges.*
2. IABSE (1984) *International Association of Bridge and Structural Engineering, 12th Congress,* Vancouver, Canada, 3–7 September.
3. BINGHAM, T.G. and LEE, D.J. (1969) The Mancunian Way elevated road structure. *Proceedings of the Institution of Civil Engineers,* **42**, 459–92.
4. ZEDERBAUM, J. (1966) The frame action of a bridge deck supported on elastic bearings. *Civil Engineering and Public Works Review,* **61** (714), 67–72.
5. LEONHARDT, F. and ANDRÄ, W. (1960) Stutzprobleme der Hochstrassenbrucken. *Beton – und Stahlbetonbau,* **55** (6), 121–32. (English translation, Library Translation Cj 95, Cement and Concrete Association, London, April 1962, pp. 30.)
6. ROARK, R.J. (1965) *Formulas for Stress and Strain,* McGraw-Hill, New York.
7. O'DONNELL, W.J. (1960) The additional deflection of a cantilever due to the elasticity of the support. *Transactions of the American Society of Mechanical Engineers, Journal of Applied Mechanics,* **27** (3), 461–4.
8. FAUPEL, J.H. (1964) *Engineering Design,* John Wiley, New York.
9. MULLER, J. (1963/4) Concrete bridges built in cantilever. *Journal of the Institution of Highway Engineers,* **10** (4), 1963, 156–71; **11** (1), 1964, 19–30.
10. DNV (1984) Buckling strength analysis of mobile offshore units. Classification Note 30.1.
11. HARVEY, W.J. (1988) Application of the mechanism analysis to masonry arches. *The Structural Engineer,* **66** (5).
12. MELBOURNE, C. (1990) The behaviour of brick arch bridges. *Proceedings of the British Masonry Society* No. 4.

THREE

Expansion joints

3.1 Introduction

Expansion joints should, more correctly, be known as movement joints since they cater for relative movement between bridge deck spans and abutments resulting from a number of causes not exclusively due to the temperature.

The expansion joint must be recognized as an integral part of any bridge structure and as such should be considered at an early stage in the design. If the expansion joint is carefully designed and detailed, properly installed by specialist operatives and given reasonable maintenance in service, there is no reason why it should not give trouble-free performance for many years.

It is important to appreciate that expansion joints are located in the most vulnerable position possible on any bridge, situated at surface level where they are subject to impact and vibration of traffic and exposed not only to the effects of natural elements such as water, dust, grit, ultra-violet rays and ozone, but also the effect of applied chemicals such as salt solutions, cement alkalis and petroleum derivatives.

To function properly, bridge expansion joints must satisfy the following conditions:

1. accommodate all movements of the structure, both horizontal and vertical;
2. withstand all applied loadings;
3. have a good riding quality without causing inconvenience or a hazard to any class of road user (including cyclists, pedestrians and animals);
4. not present a skid hazard;
5. be silent and vibration free in operation;
6. resist corrosion and withstand attack from grit and chemicals;
7. require little or no maintenance;
8. allow easy inspection, maintenance, repair or replacement.

Penetration of water, silt and grit must be effectively prevented or provision made for their removal. A survey [1] of 200 concrete bridges identified leaking expansion joints, poor or faulty drainage details, defective or ineffective waterproofing and limited access to bearing shelves for maintenance as being major factors in the deterioration of these structures.

The performance in service of many joints has been variable; the reasons for this are often not readily apparent and some previous assessments of the cause of failure have been either inaccurate or inappropriate. Surveys carried out by the Transport and Road Research Laboratory [2–4] have found performance to be influenced by the following factors:

1. structural movements of the joint;
2. traffic density and axle loading;
3. joint design;
4. materials used;
5. mix preparation, cure, compaction, shrinkage and thickness laid of surfacing and hand mixed materials;
6. bedding, bond and anchorage;
7. condition of the substrate;
8. weather and temperature during installation and service;
9. detritus, debris and corrosion;

10. site preparation and workmanship;
11. performance of the bearings.

The complex combination of factors that influenced performance was not necessarily the same combination or sequence for each individual type of joint.

Care should be taken to ensure that the design of the joint and the materials used in its construction can adequately accept the range of movements and deck end rotations that it will be called on to accommodate. Structures with relatively flexible decks can give rise to severe conditions for certain types of joint (e.g. buried joints) especially on multispan structures where the joints are subject to movements from adjacent spans. Rotation of the deck is influenced by the structural design and the efficiency of the bearings and any additional restraint of the bearings can cause increased movements at the joints, particularly with decks having a deep section. Any restriction (e.g. debris and corrosion) preventing full movement can cause severe damage to the joint and structure.

Today there are a large number of proprietary expansion joints on the market and the problem facing the engineer is often that of selecting the most suitable joint to give good performance and a trouble-free life for at least as long as that of the surfacing.

Advice on the selection, design and installation of expansion joints is given in a number of publications [4–7]. Selection of joint type is largely determined by the total range of movement to be accommodated. In multispan viaducts a limited number of joints is preferable to a large number of small ones, unless the span arrangement is such as to permit continuous surfacing over the joints.

The choice is summarized in Table 3.1 which is extracted from Department of Transport Standard BD 33/88 [5].

Table 3.1 Selection of joint type (limiting joint movement at serviceability limit state)

Joint type	Total acceptable longitudinal movement		Maximum acceptable vertical movement between two sides of joint (mm)
	Minimum[1] (mm)	Maximum (mm)	
1. Buried joint under continuous surfacing	5	20	1.3
2. Asphaltic plug joint	5	40	3
3. Nosing joint with poured sealant	5	12	3
4. Nosing with preformed compression seal	5	40	3
5. Elastomeric	5	$-^2$	3
6. Elastomeric elements in metal runners	5	$-^2$	3
7. Cantilever comb or tooth joint	25	$-^2$	3

[1]The minimum of the range is given to indicate when the type of joint may not be economical.
[2]Maximum value varies according to manufacturer or type.

3.2   Problems of geometrical layout

The expansion joint has to accommodate movements of the structure in the longitudinal, transverse, vertical and rotational modes. Analytical methods of assessing the displacements and forces associated with this structural behaviour have already been discussed, but there are several features, mainly concerning the geometry of the structure and the joint, which require careful consideration if trouble-free operation is to be achieved.

With structures curved in plan or with skew joints, the relative movement may not be normal to the line of the joints (Figs 3.1 and 3.2). This can lead to binding of the elements or high shear forces in filler materials which may be exuded and carried off by the traffic, ultimately causing failure. It is therefore important to assess this transverse displacement and to design the joint accordingly, or eliminate the movement by restraining the bridge either at the joint or preferably at the bearings.

A similar type of problem may be experienced in a vertical direction through excessive flexural rotation at the joint (Fig. 3.3) but, by arranging the end bearing and expansion joint in the same vertical plane, the discontinuity can be minimized and the motion reduced to a purely horizontal displacement. On joints designed for small movements, this may also be overcome by providing an articulated running-on slab.

As it is usual to set the bearings horizontally a discontinuity may be introduced at the expansion joint if the structure is on a vertical grate Θ (Fig. 3.4). This problem is significant if the total movement X is large because, with the joint set at grade for the mean position, the magnitude of the vertical discontinuity is given by $(X \tan \Theta)/2$. There may be a sudden jump or a change of grade over a certain length depending on the type of joint. The problem can be overcome by sloping the bearings if the supporting structure can withstand a force applied in the same direction as the upward slope. Alternatively, a compensating mechanism such as a wedging device may eliminate or reduce the effect to an acceptable magnitude but this usually means added complication and expense.

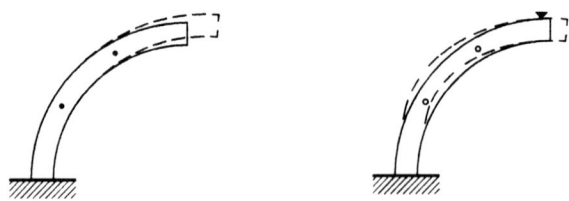

Fig. 3.1   Unrestrained and restrained movement of a curved structure.

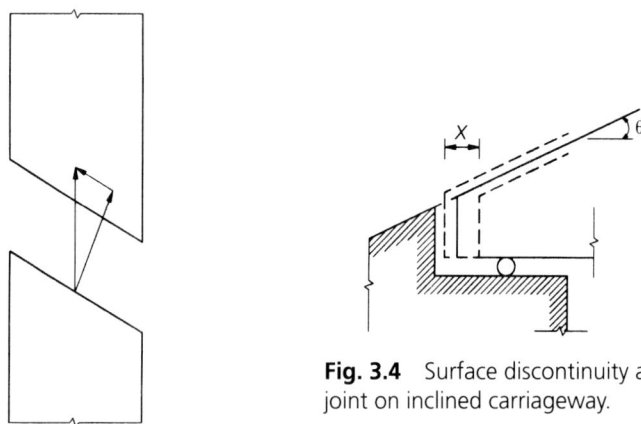

Fig. 3.4   Surface discontinuity at joint on inclined carriageway.

Fig. 3.2   Normal and shear displacements across skew joint affecting filler material.

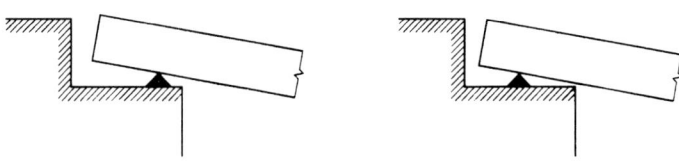

Fig. 3.3   Position of bearing affects discontinuity of joint due to end rotation.

3.3   Joints for small movements

For movement of less that 5 mm (0.2 in) it is usually considered that no special provision is necessary, and for movements up to about 10 mm (0.4 in) the most common treatment is for a gap-filled or buried joint with continuous surfacing, the filling being protected by the surfacing which also absorbs much of the impact. As water usually percolates through the road surfacing, the joint should be made to a natural fall wherever possible to assist drainage. Even when flashing is present the joint itself will rarely remain waterproof, so it is essential to make provision for effective drainage under the joint to prevent straining and deterioration of abutments and columns.

On bridges employing a continuous membrane over the deck area, particular care is required to maintain continuity of the membrane over the joint. The waterproof layer can be plastic, bituminous felt or copper membrane, although sprayed liquid treatments are employed. A typical detail is shown in Fig. 3.5 where recesses are left in the mating units to accommodate a plate to support the surfacing and ensure a smooth profile across the joint. Various materials have been used to form buried joints, e.g. copper-lined bituminous sheeting, quarry tiles, melamine sheet or steel plate with varying degrees of success depending upon the horizontal movements and loading applied. The main problem with rigid plates is the difficulty of bedding them down so that they do not rock under traffic loading and cause premature breakdown of the joint.

For movements up to 10 mm a proprietary flashing may be appropriate provided there is a minimum of 100 mm surfacing. For movements of 10–20 mm an elastomeric pad may be installed on top of the flashing to support the surfacing. It has been found [4] that the use of a sliding layer faced on both sides with aluminium foil improves performance.

Buried joints are more prone to failure resulting from traffic induced movements than are other types of joint. This is particularly so where the material spanning the expansion gap does not spread movement over an appreciable length of deck. The performance of buried joints can be improved by reinforcing the surfacing over the joint with

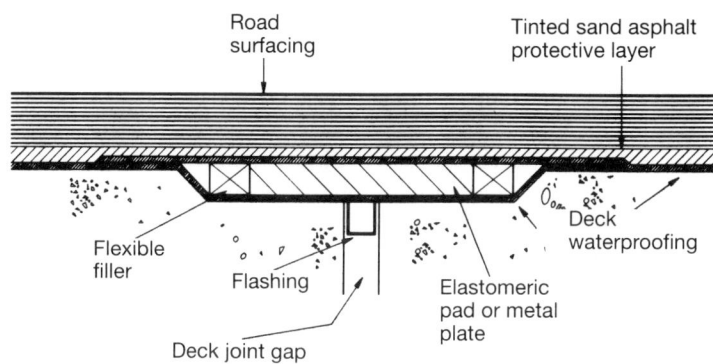

Fig. 3.5   Buried deck expansion joint for ±10 mm movement.

expanded metal. The use of rubberized polymer asphalt surfacing laid as part of the joint improves its flexibility and hence durability. Hot-rolled asphaltic surfacings should be at least 120 mm thick for long term durability, but for the more flexible asphalts this figure can be reduced to 85 mm.

3.4   Joints for medium movements

Asphaltic plug joints are a variation on the buried joint. There are a number of proprietary joint systems included in this description. The joint is usually constructed in layers using a mixture of flexible material and aggregate to provide not only the homogeneous expansion medium, but also the running surface at carriageway level (see Figs 3.6–3.8). The binders used are usually based on bitumens modified with plasticizers and polymers to obtain the desired flexibility. The coarse aggregate used within the joint material is usually of the basalt group. The plate across the expansion gap is essential in preventing the extrusion of the flexible joint material into the gap under axle loading. It may be fitted with a vertical lug to assist with location.

The system was developed during the 1970s and was used initially to cater for small movements. However, although the system coped successfully with these movements, in some cases the joint material was too flexible and suffered from tracking and flowing especially during hot weather. The system was improved by increasing the density and stiffness of the material, mainly in the top layers up to carriageway level. In general, asphaltic plug joints are now formulated to work satisfactorily in the movement range given in Table 3.1 provided the adjacent surfacing is not less than 100 mm thick, the gradients and crossfalls are not too severe and the bridge deck is not noticeably lively at the joints.

The low initial capital cost and speed of installation, combined with a belief that these joints have a greater tolerance of installation procedures and site conditions, are considered by many authorities to be particularly advantageous. These benefits are particularly relevant with respect to the time constraints imposed by lane rental contracts. However, reservations have been expressed by some authorities regarding the performance of these joints and it has been suggested that a more rigorous specification control should be imposed. The performance of asphaltic plug joints is sensitive to the standard of workmanship during installation.

Tests carried out in Britain have shown that carriageway expansion joints with gaps up to 80 mm in width do not have an adverse effect on

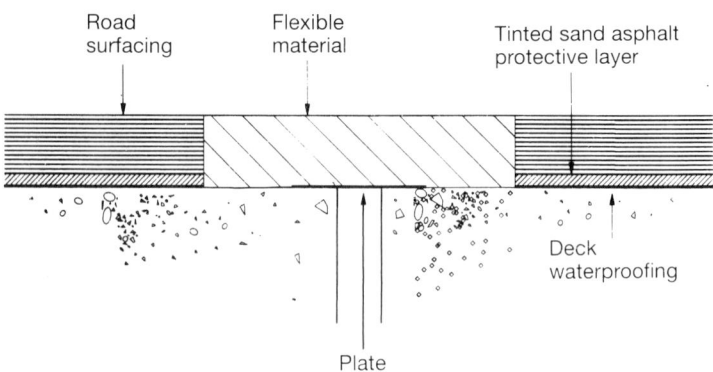

Fig. 3.6   Asphaltic plug expansion joint for 15 mm to 45 mm movement.

Fig. 3.7   Completed Thorma-Joints. (Courtesy Prismo Limited.)

riding comfort. For movements up to about 50 mm (2.0 in) the most popular type of joint during the 1960s and 1970s was the preformed flexible sealing strip compressed between nosings (Figs 3.9–3.11). These joints are suitable where pedestrian, cycle or animal traffic has to be accommodated as they provide a continuous surface. However, the engineer should ensure that arrangement for accommodating kerbs, edge beams and medians are adequate as it is often at these points that trouble starts.

A survey of nosing joints in the UK [3] has shown that their effectiveness is dependent on the nature of the design and materials, workmanship and condition of the substrate at installation. Rigid mix materials for nosings were prone to cracking and bond failures were more prevalent where there was no mechanical tie or the sealant became incompressible. Agents such as rubber, polysulphides or pitch which increase the flexibility of the nosing tend to reduce cracking but may not be compatible with the gap sealant. Alumina-based aggregates for resinous mixes tend to reduce shrinkage and have been found to give long-term benefits, particularly in combination with flexible mixes. Chopped wire reinforcement added to cementitious mixes helps reduce cracking.

Most of the cellular fillers are based on rubber or neoprene, although foamed and expanded plastics are used. Neither solid rubber or neoprene nor expanded neoprene are now recommended, as the former are expensive in relation to the small permissible compressive strains and the latter appears to suffer loss of elasticity at low temperature and with time. Although the normal wearing properties are good, they are subjected to severe treatment and prone to damage; provision for easy maintenance is therefore essential. It is the author's opinion that these joints should be conservatively regarded as gritproof rather than waterproof. It is therefore prudent to provide at least elementary drainage on the underside and to arrange surface slopes and gully positions so as to prevent as much water as possible from reaching the joint.

These joints require planned routine inspection if they are to perform satisfactorily, and any large deposit of silt and grit, particularly around kerb details, should be removed, because it can severely restrict the free movement and may support vegetation growth which can damage the sealer.

Fig. 3.8 Thorma-Joint binder expansion test. (Courtesy Prismo Limited.)

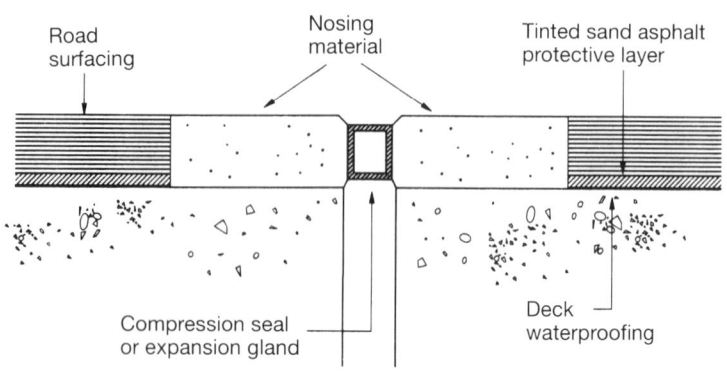

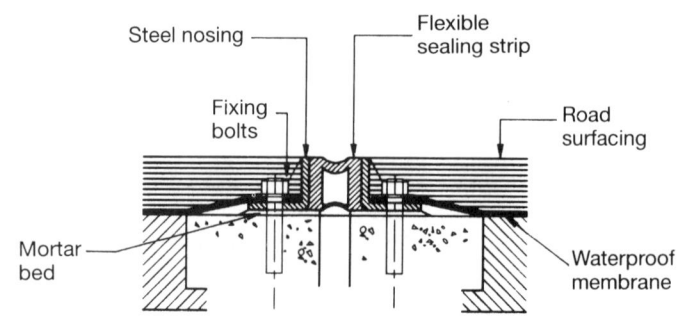

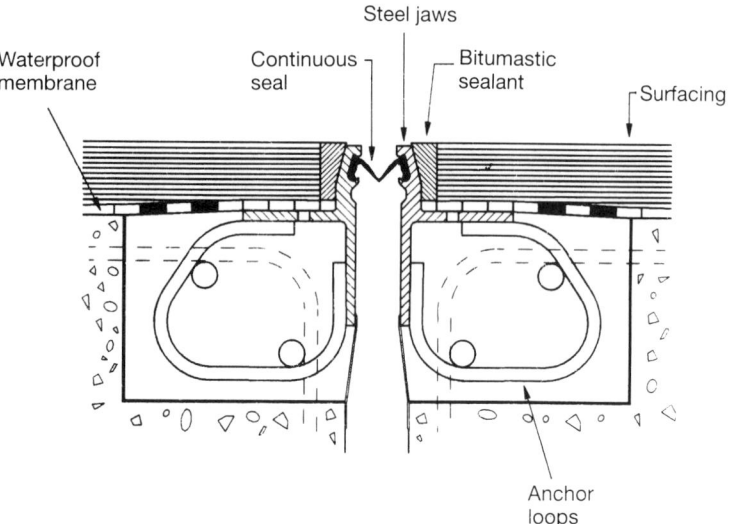

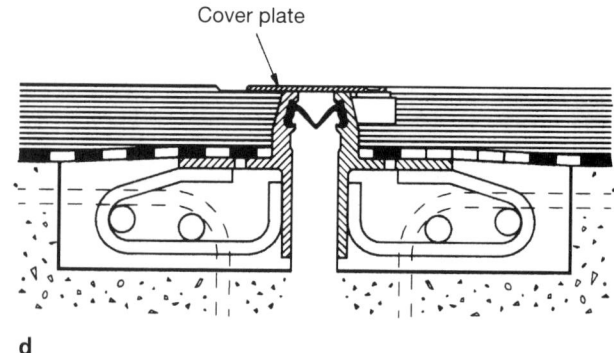

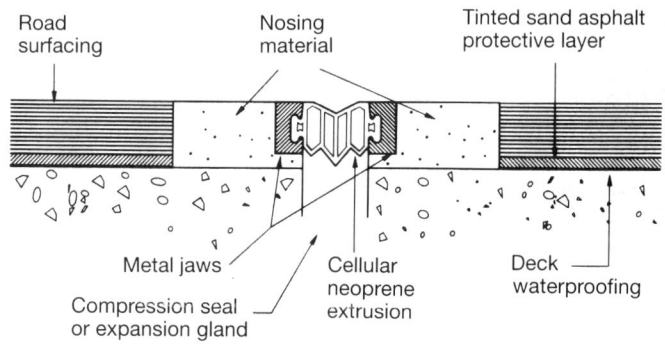

Fig. 3.9　Preformed sealing strip joint for 15 mm to 50 mm movement, (a) epoxy nosings; (b) steel nosings; (c) v-joint has 20 mm to 80 mm movement; (d) typical footpath detail; (e) cellular joint.

a

b

Fig. 3.10 Britflex joint. (Courtesy Britflex Limited), (a) detail;
(b) completed joint.

In using precompressed sections the forces exerted on the structure by the sealer should not be ignored; the modulus of poured sealants and the hollow tube elastomers is usually of the order of 170 N per millimetre compression per metre length. For example, a sealer compressed 25 mm (1 in) over a 15 m (50 ft) wide bridge could exert a force of some 64 kN (6.5 tonf) at surface level of the structure. Manufacturers of such products often do not quote the modulus of their products in their sales literature. Since this is a surface-level joint, the fixing arrangement requires careful detailing and skilled installation on site.

During the early 1970s epoxy mortar nosings were popular due to their relative cheapness and ease of maintenance, but these have not always stood up well in service, the deterioration usually being attributable to poor materials and workmanship at installation. In spite of improvements in the formulations of epoxy nosings, which have increased their success, they have been superseded to some extent by cementitious, polyurethane and polyureide binders, which are more tolerant of adverse site conditions and have a better success rate in service.

Very many epoxy nosings have been replaced. One material favoured for this purpose is Monojoint HAC, a cement-based nosing material incorporating wire fibre reinforcement. This effectively eliminates the two major factors thought to contribute to the failure of epoxy nosings; i.e. the differing coefficients of expansion of concrete and epoxy mortar causing the shear forces on the bond plane and the exothermic action of epoxy mortar under cure producing shrinkage stresses. Another material used for replacing epoxy nosing joints and repairing damaged buried joints is Thorma-joint. This is an asphaltic plug type joint and consists of a combination of single-sized roadstone aggregate and a specially formulated rubberized bitumen compound. Reports indicate that this material is standing up well under traffic conditions.

If steel sections are used to form the nosings they should be of robust construction since they are required to take severe impact loads from traffic. Many joints have been proved unsatisfactory because of failure of the fixings and, where angles are used to reinforce the opposing edges of the structure, care should be taken to achieve adequate

a

c

b

d

: angles.
to the
~~d

also advisable to
\g-down arrange-
cast-in anchorage
lds.

b preformed steel
lodated (see Fig.
lloroprene rubber,
\g in steel nosing
the use of suitable
le jaws are rigidly
reinforcing steel
. Such joints are
2 and 3.13). The
eoprene extrusion
rcement cast into
). However, there
ld hence the jaws)
bond between the

plates to enable the joint to span the expansion gap without deflecting significantly under load. Movement is accommodated by shearing strains in the elastomeric material. The joint is bolted down flush with the wearing surface and, if manufactured in one continuous piece, should be waterproof, although care must be taken with the details at kerbs and edges. To reduce the hazard of skidding prevalent with wet rubber, some anti-skid treatment, usually grooving of the top surface, is applied. However, reinforced elastomeric points are often supplied in discrete lengths with butt joints parallel to the line of traffic. Although vulcanizing the elastomer to provide a seal is possible, it is difficult to achieve on site and a waterproof membrane is essential to prevent leakage.

Major problems have been encountered in maintaining a watertight seal with these joints where elements are bolted down. This appears to be mainly due to relaxation and a tendency for the holding-down bolts to unscrew due to traffic-induced vibration. Unevenness and excessive noise can be a problem with these joints.

A variant of the joint incorporating a preformed seal comprises a flexible gland or strip of reinforced neoprene set in nosing blocks of solid neoprene, reinforced with steel plate, which are bolted down to the bridge deck and abutment structures.

A further variant is the PSC Freyssi expansion joint which has a continuous cellular elastomeric extrusion clamped between an upper steel plate and a lower casting which is bolted down to the bridge deck. This allows the expansion membrane to be easily replaced.

An alternative to the cellular preformed sealer consists of a shaped slab of neoprene (Figs 3.14 and 3.15) usually reinforced with steel

Fig. 3.11   Britflex joint installation on M6 motorway. (Courtesy Britflex Limited), (a) setting temporary spacer plates for correct opening; (b) welding kerb section to main joint section; (c) pouring Britflex jointing compound into recess (note spiral reinforcement welded to runners); (d) inserting sealing strip into nosing rails.

Fig. 3.12 Mageba expansion joint. (Courtesy Mageba S.A.), (a) detail of reinforcement; (b) installation of neoprene seal; (c) kerb upstand detail.

Fig. 3.13 Mageba expansion joints and bearings were used on the Tavernauto Road Viaduct at Kärhten, Austria.

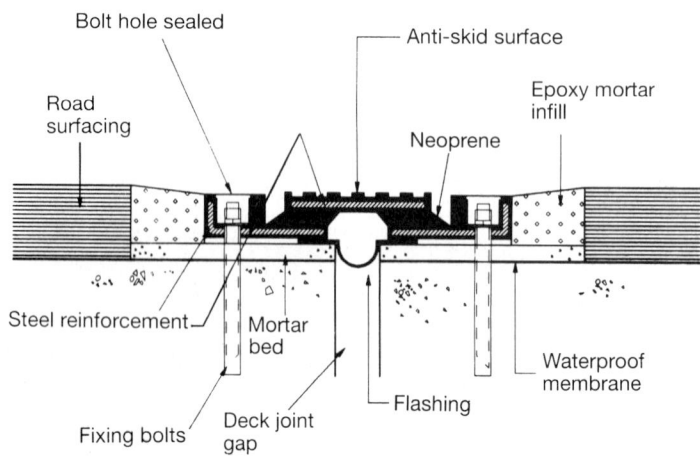

Fig. 3.14 Elastomeric cushion joint for 50–300 mm movement.

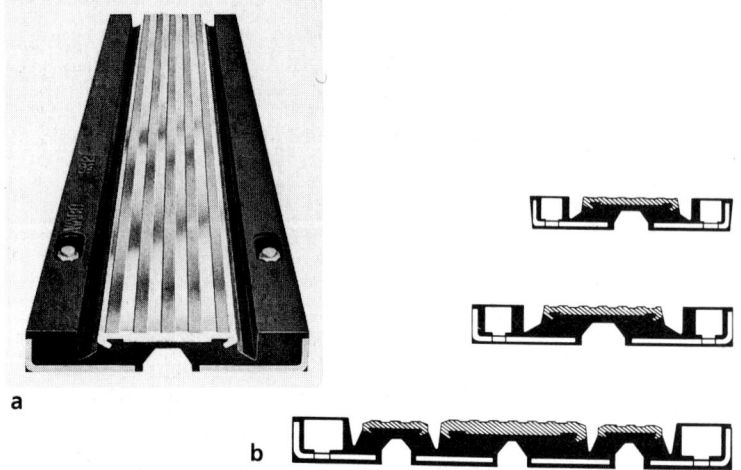

Fig. 3.15 Waboflex SR joints. (Courtesy Servicised Ltd), (a) general view; (b) sections; (c) joint installed on Ware bypass in Hertfordshire; (d) Waboflex SR-13 upstand kerb detail incorporating factory welded and vulcanized fabrication.

3.5   Joints for large movements

Joints for large movements are often of the open type using sliding plates, cantilever or propped cantilever tooth or comb blocks (Figs 3.16 and 3.17). It may not be practical to seal joints where the total movement exceeds 50 mm (2 in) and adequate provision must therefore be made for the disposal of surface water, grit, salt, etc., with easy access for maintenance. Because of the passage of surface water through the joint and splashing in and around the collector system, it is vital that adequate protective treatment should be applied to any parts of the joint exposed to these corrosive elements. The drainage system should be constructed to generous falls as a means of self-cleaning.

With joints of this type, there is always a danger of the free movement being restricted by a build-up of grit and other non-compressible matter between the moving elements [8]. Very often, vegetable growth near the kerbs further consolidates the intrusion and effectively impedes the proper functioning of the joint, imparting high and unexpected loads to the structure, often with disastrous results to both structure and joint.

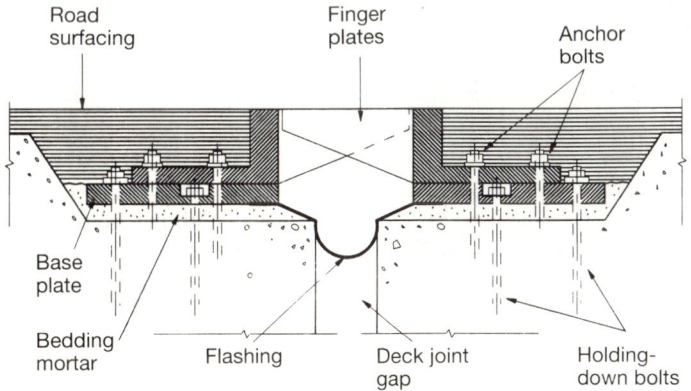

Fig. 3.16   Cantilever comb or toothed joint.

Fig. 3.17   Westway expansion joint after 20 years in service. (Courtesy G. Maunsell and Partners.)

a

bridges and continuous girder bridges. The link type expansion joint consists of interlocking finger plates supported by cross beams at each end and in the centre of the joint. The fingers are fixed to the end cross beams and slide on the central beam which is supported by a linkage system which transfers the load back to the ends. Roller bearings take the vertical loads at the ends while the joint is fixed to the structure through universal joints which take only horizontal loads (Fig. 3.24). The link mechanism also keeps the three cross beams at equal spacing. The finger plates are grooved to prevent vehicle slippage. The joint is made in sections across the bridge deck, each one being fixed to the bridge through only one pair of universal joints allowing it to respond to all possible displacements of the bridge. A cover plate version is available for footpaths.

b

Fig. 3.23 Roller shutter joints, (a) Glacier roller shutter joint. (Courtesy The Glacier Metal Co. Ltd); (b) demag roller shutter joint, Westway, London, after 20 years in service. (Courtesy G. Maunsell and Partners.)

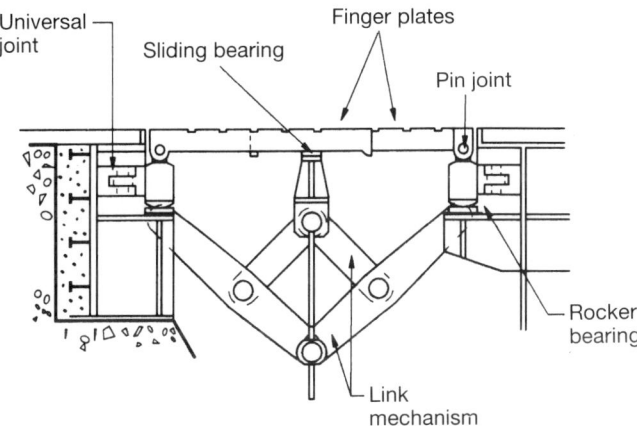

Universal joint
Sliding bearing
Finger plates
Pin joint
Rocker bearing
Link mechanism

Fig. 3.24 Link expansion joint.

3.6 Longitudinal joints

It is not unusual for bridge decks to be constructed with a longitudinal joint separating the two carriageway decks. In some instances the decks are separated so that the detail of the edge beam along the separation does not differ from the outside edge and is treated as such for water disposal. However, there are bridges where the two carriageways are separated by a narrow gap and a barrier common to both carriageways prevents vehicles crossing the central reservation. This gap can be subject to considerable differential vertical movement.

Road salt and spray inevitably find their way down these open joints causing considerable contamination of the concrete by chloride ions leading to reinforcement corrosion, particularly if the two separate carriageways are supported on a common substructure. The large differential vertical movement precludes the use of the normal type of joint seal and special designs have to be adopted. For example the type of joint adopted to overcome the large differential movement of up to 200 mm and considerable ingress of chloride on the Tees Viaduct in Cleveland, UK incorporated a flexible seal manufactured from material consisting of nylon fabric coated both sides with chlorosulphonated polyethylene synthetic rubber. This material was manufactured to specified breaking, tear and adhesion strengths. The seal, looped in section, was clipped into plastic clasps bonded to the concrete and formed a unit which was shown to produce an effective seal. The sealing membrane must be suitably stiffened to prevent debris collecting in hollows causing unnecessary wear. It is important that the seal is properly bedded and particular attention must be paid to joints between lengths of seal to ensure that they are watertight.

3.7   Design requirements

In the UK the Department of Transport Standard BD 33/88 [5] lays down that the design of expansion joints shall be such that the joints will function correctly without the need for excessive maintenance during their working lives. In addition, the joint and its installation shall be capable of withstanding the ultimate design loads and movement range given in the Standard, i.e. $R^* > S^*$ where R^* is the design resistance and S^* is the design load effect. The design load effect S^* is determined by multiplying the effects of the design loads and movements by γ_{f3}. γ_{f3} is taken as 1.0 for the serviceability limit state and 1.1 for the ultimate limit state.

The design resistance of the various parts shall be in accordance with BS 5400: Part 3 for steel elements and BS 5400: Part 4 for reinforced concrete components as implemented by BD 13/90 [11] and BD 24/84 [12]. Elastomeric elements need only be designed for the serviceability limit state. Components, such as anchor bolts subject to fluctuating loads shall be checked for fatigue in accordance with BS 5400: Part 10 as implemented by BD 9/81 [13].

BD 33/88 [5] states that the nominal vertical load shall be taken as either a single wheel load of 1000 kN or a 200 kN axle with a 1.8 m track. The load from each wheel shall be applied such as to give an effective pressure of 1.1 N/mm^2 over a circular area of 340 mm diameter. The load shall be applied separately to either edge of the joint to give the most severe effect. The nominal horizontal traffic load shall be taken as a uniformly distributed load of 80 kN/m run of joint acting at right angles to the joint at carriageway level. This is combined with any loads resulting from strain of the joint filler or seal over the nominal range of movements. These include temperature movement, creep and shrinkage, lateral movement and settlement. These nominal loads are multiplied by a partial load factor γ_{fL} to give the design load effect. BD 38/88 gives the following values for γ_{fL}:

	Wheel loads	Horizontal load
Serviceability limit state	1.20	1.00
Ultimate limit state	1.50	1.25

3.8   Subsurface drainage

Water trapped within the road surfacing on the high side of a deck joint can, through hydraulic pressure from wheel loading, cause failure of the bond or seal between the joint and the waterproofing systems. This may result in water leakage into the deck joint gap and hence onto the adjacent concrete. To prevent this from occurring a subsurface drainage system should be provided as shown in Figs 3.25 and 3.26, which illustrate some of the forms that subsurface drainage may take.

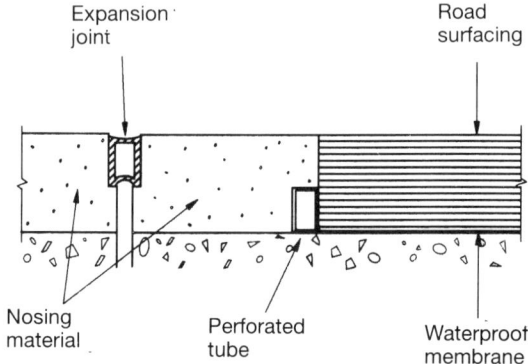

Fig. 3.25   Subsurface drainage (perforated tube).

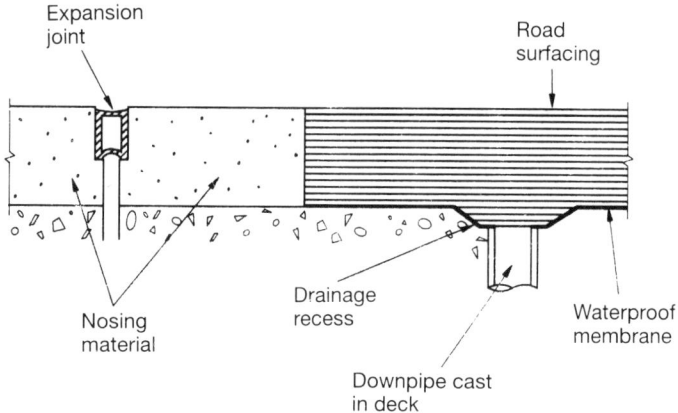

Fig. 3.26   Subsurface drainage (drainage recess).

Figure 3.25 is typical of current practice in France and Germany and certain UK systems. The buried galvanized steel perforated drain tube may be either circular or rectangular in cross-section and discharges water via a suitable connection to the bridge drainage system. It has been found that narrow slots tend to become clogged with silt and current practice is to use holes of about 10 mm diameter at 100 mm centres.

Figure 3.26 shows another detail developed in Italy. The down pipes are spaced at intervals to suit the bearing and jacking point positions. The trough is filled with surfacing which is sufficiently permeable to permit the passage of water. The pipes should be carefully positioned in order that drips from the outlets do not damage the face of adjacent concrete.

3.9   Rail expansion joints

There are two types of expansion joints used with rail track to cater for movement at bridge expansion joints. One, known as the British Rail (BR) type, although it originated in Belgium, consists essentially of a parallel-sided scarved joint in the rail (i.e. the rail is split longitudinally with one half sliding over the other) (Figs 3.27 and 3.28). The normal BR scarved joint, which is under 1 m long, caters for a maximum 130 mm of movement. During passage over the joint the wheel load is on only one half the rail width leading to high stresses. In the normal case of long welded rail fixed to concrete sleepers on ballast the rail only expands over the end 30 m. The adjustment switch should be well anchored by self-weight and located well away (5–10 m) from the point of fixing of the bridge expansion joint.

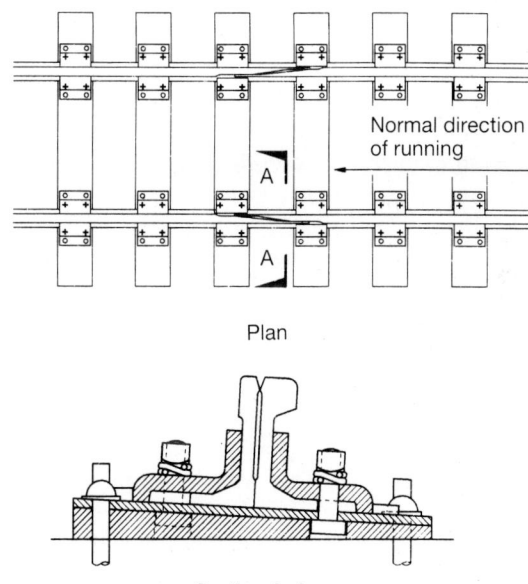

Plan

Section A–A

Fig. 3.27   Parallel adjustment switch (BR type).

Fig. 3.28   British Rail expansion joint. (Courtesy R.G. Bristow.)

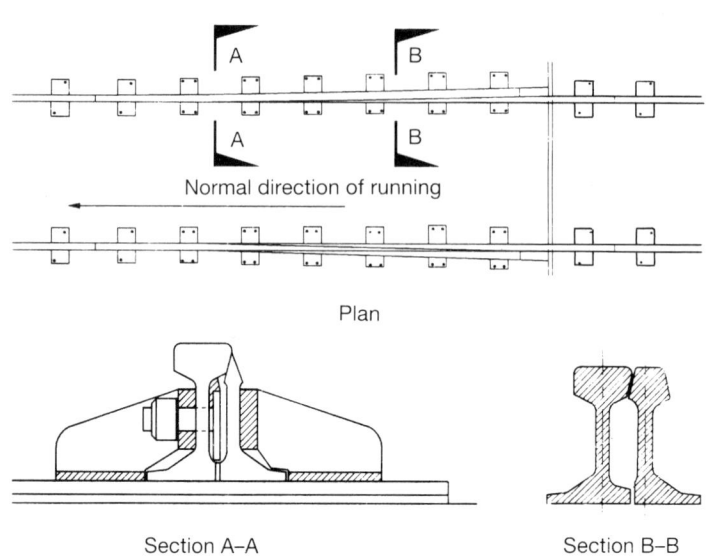

Plan

Section A–A   Section B–B

Fig. 3.29   Tapered expansion switch (LT type).

Account must be taken of the rail breathing on the landward side of the joint and allowance for this must be included in the expansion rail joint as well as movement on the bridge. There is a Belgian double version of this joint which allows movement on the bridge side and the landward side. Because rail track is usually laid on ballast, which allows some movement of the track relative to the bridge, British Rail's experience indicates that bridge movements are not as great as expected and it has never been found necessary to employ other than their standard 130 mm joint.

The other version is a long tapered scarved expansion switch (Figs 3.29 and 3.30) (sometimes known as the London Transport (LT) type since it was their standard for expansion joints, as until recently LT only employed bullhead rail). This was the original version of the expansion joint and is suitable for quite large movements. It should only be used in the trailing direction, i.e. trains should not run towards the pointed rail end. A 1 in 70 machined side slope is usual; a flatter slope makes the joint longer and more difficult to hold down, since the sliding part is held only by clamps, and hence requires more

Fig. 3.30   Long tapered expansion switch. (Courtesy R.G. Bristow.)

maintenance. This type of joint gives rise to gauge widening on expansion; up to 5 mm with a 1 in 70 side slope. This can cause rough riding with high speed trains. British Rail no longer use them; they rely on normal long welded rail expansion joints each end of a bridge. They have found that the track tends to sort itself out under vibration caused by the passage of trains. Having the pointed end move makes for simpler design. The adjustment switch should not be located directly over the bridge expansion joint as this will cause bending in the rail.

References

1. Department of Transport (1989) The Performance of Concrete in Bridges: a survey of 200 highway bridges. HMSO, London
2. PRICE, A.R. (1982) The service performance of fifty buried type expansion joints. TRRL Report SR 740, Transport and Road Research Laboratory, Crowthorne.
3. PRICE, A.R. (1983) The performance of nosing type bridge deck expansion joints. TRRL Report LR 1071, Transport and Road Research Laboratory Crowthorne.
4. PRICE, A.R. (1984) The performance in service of bridge expansion joints. TRRL Report LR 1104, Transport and Road Research Laboratory, Crowthorne.
5. Department of Transport (1989) Expansion joints for use in highway bridge decks. Departmental Standard BD 33/88.
6. Department of Transport (1989) Expansion joints for use in highway bridge decks. Departmental Advice Note BA 26/88.
7. KOSTER W. (1969) *Expansion Joints in Bridges and Concrete Roads*. Maclaren and Sons.
8. BUSCH, G.A. (1986) A review of design practice and performance of finger joints. Paper presented to the 2nd World Congress on Joint Sealing and Bearing Systems for Concrete Structures, San Antonio, Texas, September.
9. WATSON, S.C. (1972) A review of past performance and some new considerations in the bridge expansion joint scene. Paper presented to regional meetings of the AASHO Committee on Bridges and Structures, Spring.
10. KOSTER W. (1986) The principle of elasticity for expansion joints. Paper presented to 2nd World Congress on Joint Sealing and Bearing Systems for Concrete Structures, San Antonio, Texas, September.
11. Department of Transport (1982) Design of steel bridges, use of BS 5400: Part 3: 1982. Departmental Standard BD 13/90.
12. Department of Transport (1984) Design of concrete bridges, use of BS 5400: Part 4: 1984. Departmental Standard BD 24/84.
13. Department of Transport (1981) Code of practice for fatigue, implementation of BS 5400: Part 10: 1980. Departmental Standard BD 9/81.

FOUR

Bridge bearings

4.1 Introduction

Until the publication of Part 9 of BS 5400 [1] dealing with the design, manufacture, testing and installation of bearings for steel, concrete and composite bridges, there had been no British Standard dealing comprehensively with bridge bearings. General guidance on modern bearings was available in certain publications [2,3] and information on roller, rocker and metal-to-metal sliding bearings was given in early textbooks [4] and papers [5].

Up to the middle of this century bridges relied on roller, rocker or metal sliding bearings to permit movement. With more advanced designs to make better use of the materials employed and the increased use of skewed and curved bridges to carry modern high speed roads over obstructions, the need arose for bearings to take movement in more than one direction. New types of bearings have been developed taking advantage of the new materials arising from improved technology. No doubt others will be developed in the future, but it will be necessary to ensure that they are at least as reliable as those already in service.

BS 5400: Part 9 [1] does not cover concrete hinges and link bearings nor bearings for moving bridges (e.g. swing and lift bridges). Also it does not include bearings made with proprietary materials such as Fabreeka and Bonafy but provision is made for the use of such materials, provided the engineer is satisfied as to their long-term suitability for the function intended. The document is split into two sections: Part 9.1 is a Code of Practice and gives rules for the design of bearings. Part 9.2 specifies the materials, method of manufacture and installation of bearings.

4.2 Function of bearings

Bridge bearings fulfil a number of functions. They are:
1. to transfer forces from one part of the bridge to another, usually from the superstructure to the substructure;
2. to allow movement (translation along, and/or rotation about any set of axes) of one part of a bridge in relation to another;
3. by allowing free movement in some directions but not in others, to constrain that part of the bridge supported by the bearings to defined positions and/or directions.

Constraint provided in (3) above gives rise to forces which have to be transmitted by the bearing to the supporting structure. The freedom of movement permitted by any bearing is, in practice, always affected by friction.

To achieve the required degrees of freedom of movement a complete bearing usually consists of several elements, each permitting a particular movement, the sum of which is the total freedom required. To some extent all movements can be accommodated by elastomeric bearings.

4.3 Types of bearing (Fig. 4.1)

4.3.1 ROLLER BEARINGS

Roller bearings consist essentially of one or more steel cylinders between parallel upper and lower steel plates. Gearing or some other form of guidance should be provided to ensure that the axis of the roller is maintained in the desired orientation during the life of the bearing. Roller bearings with a single cylinder can permit rotation about a horizontal axis and translation parallel to a perpendicular axis. Multiple cylinders require another element such as a rocker or knuckle bearing to permit rotation. Multiple roller bearings consist of a number of cylinders between pairs of plates and permit higher loads to be taken. Two level roller bearings consist of cylinders between three parallel plates and can be designed to permit multidirectional horizontal movements. Multiple roller bearings are little used these days having been supplanted by curved sliding or pot bearings.

4.3.2 ROCKER BEARINGS

Rocker bearings consist essentially of a curved surface in contact with a flat or curved surface and constrained to prevent relative horizontal movement. The curved surface may be cylindrical or spherical to permit rotation about one or more axes. Rocker bearings on their own will not permit translation and are normally used at the fixed end of a bridge to complement roller bearings. Rocker bearings permit rotation by rolling of one part on another.

4.3.3 KNUCKLE PIN BEARINGS

Knuckle pin bearings consist of a steel pin housed between an upper and lower support or trunnion member each having a curved surface which mates with the pin. Lateral loads are transmitted by flanges on the ends of the pin. Knuckle bearings permit rotation by sliding of one part on another. This type of bearing is seldom used today.

4.3.4 LEAF BEARINGS

Leaf bearings consist essentially of a pin passing through a number of interleaved plates fixed alternatively to the upper and lower bearing plates. Pin bearings permit only rotational movement, but can be used in conjunction with roller bearings to provide rotation and translation. Leaf bearings can be designed to resist uplift. This type of bearing is seldom used today.

4.3.5 LINK BEARINGS

Link bearings consist of plate, rod, I or tubular section members connected at their ends by pins to opposite sides of a joint to transmit vertical shear or a bearing reaction. They permit rotation and longitudinal movement by sway about a vertical axis. They are unsuitable for carrying transverse loads and are normally used in conjunction with a lateral restraint bearing free to slide vertically and longitudinally and positioned at the deck centre line. Plate or rod links are used in tension while I or tube section are suitable for use in compression or where the load can be either tensile or compressive.

4.3.6 SLIDING BEARINGS

Sliding bearings consist essentially of two surfaces of similar or dissimilar material sliding one on the other. The surfaces can be plane or curved. Plane surfaces permit translation. Curved surfaces may be cylindrical or spherical for uniaxial or multiaxial rotations respectively, and provide restraint against translation.

4.3.7 POT BEARINGS

Pot bearings consist essentially of a metal piston supported by a disc of unreinforced elastomer of relatively thin section which is confined within a metal cylinder. As the elastomer is fully confined within a metal cylinder, it provides a load carrying medium whilst at the same time providing the bearing with a multidirectional rotational capacity. By themselves, pot bearings will not permit translation.

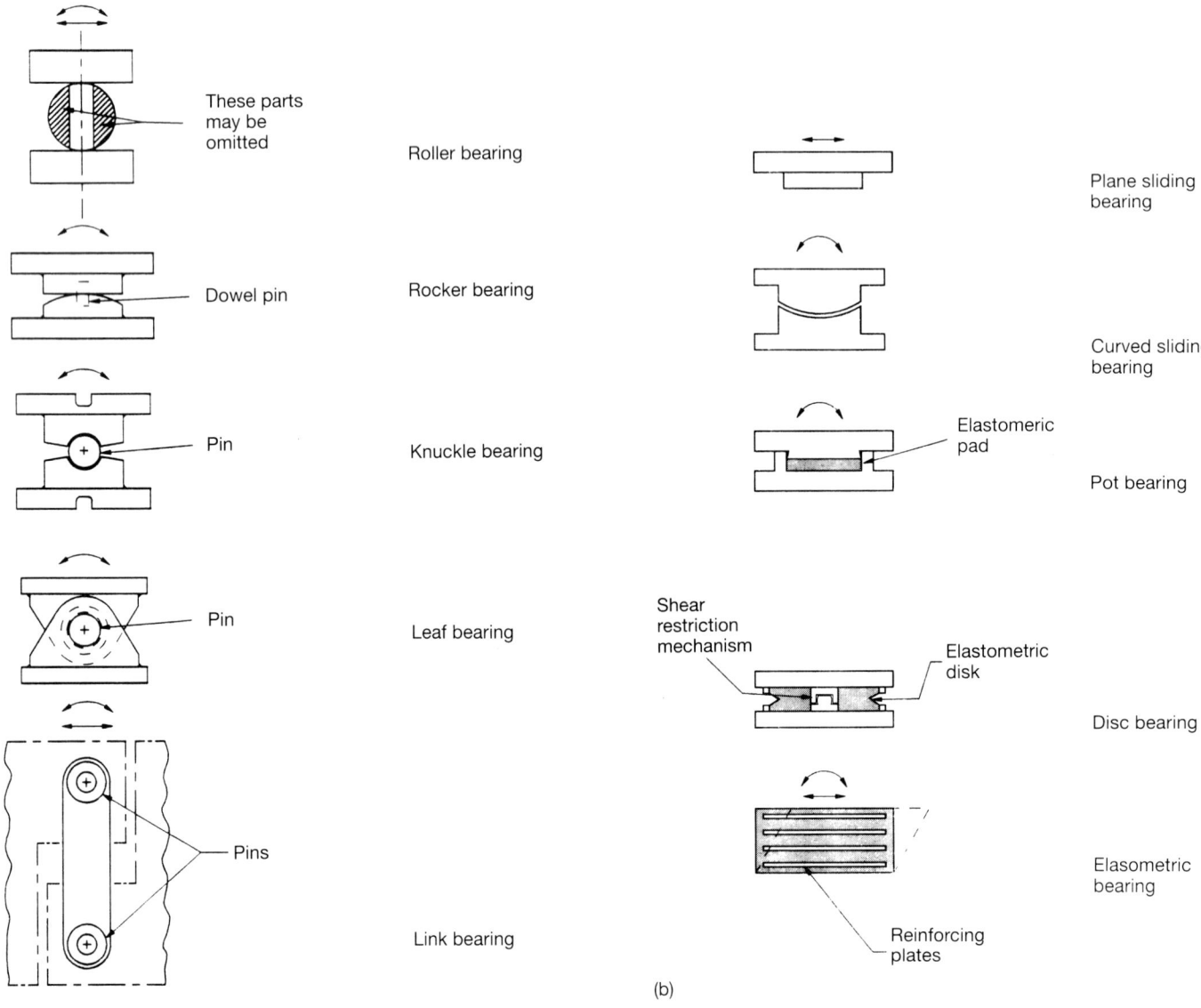

(a)

(b)

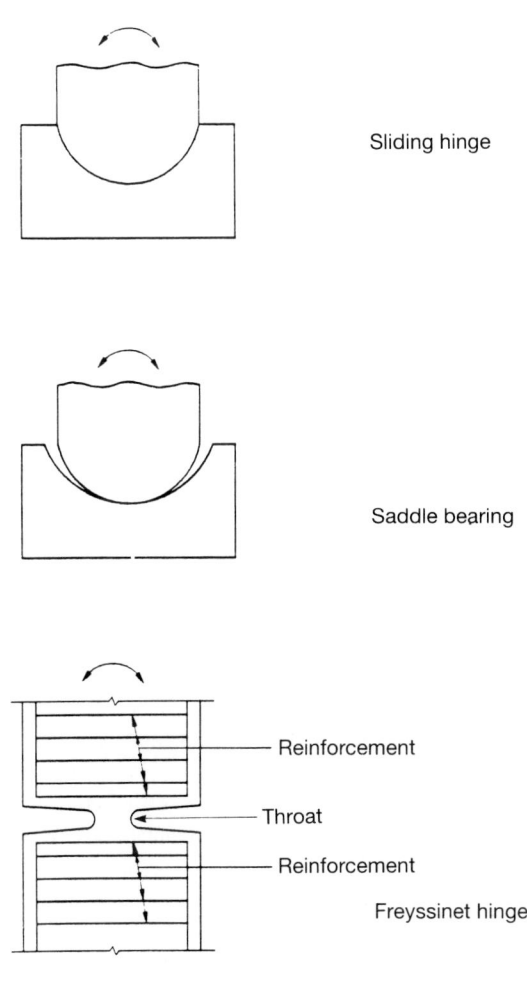

Sliding hinge

Saddle bearing

Reinforcement

Throat

Reinforcement

Freyssinet hinge

(c)

Fig. 4.1 Characteristic bearing arrangements.

4.3.8 DISC BEARINGS

Disc bearings consist of an unconfined polyether urethane disc to accommodate rotation located between two metal plates. Horizontal shear forces are transmitted between upper and lower plates by a dowel pin mechanism in the centre of the bearing. By themselves disc bearings will not permit translation.

4.3.9 ELASTOMERIC BEARINGS

Elastomeric bearings are available in two basic forms:
1. laminated bearings comprising one or more slabs of elastomer bonded to metal plates in sandwich form;
2. bearing pads which are single unreinforced pads of elastomer (i.e. without metal plates) of relatively thin section. A bearing strip is a continuous bearing pad for which the ratio of the length parallel to the abutment to the breadth parallel to the span of the bridge exceeds 10.

Translational movement is accommodated by shear in the elastomer, one surface of which moves relative to the other. Rotational movement is accommodated by the variation in compressive strain across the elastomer.

The elastomer should have sufficient shearing flexibility to avoid transmitting high horizontal loads and sufficient rotational capacity to avoid transmission of significant moments to the supports. The vertical stiffness should be such that significant changes in height under loads are avoided.

The elastomer would normally be natural or synthetic rubber.

4.3.10 CONCRETE HINGES

Concrete hinges (Fig. 4.42) perform the functions as defined above for bearings. They can take one of four forms:
1. Sliding hinges in which a convex cylindrical surface registers in a cylindrical concavity of the same radius. The surfaces may be lined with a suitable medium to reduce friction. If friction

between the two surfaces is large the moment transmitted may be significant, no rotation being possible until this friction is overcome.

2. Saddle bearings are similar to sliding hinges except that the radii of the surfaces are unequal, thereby causing a rolling rather than pivoting action. There must be sufficient friction to prevent slip between the surfaces. Saddle bearings are used when it is required to accommodate high rotations and shear forces in conjunction with relatively low axial forces.

3. Mesnager hinges, which consists fundamentally of reinforcement bars crossing each other in a small gap between the members to be articulated. Generally the bars are protected against corrosion by mortar or concrete but the contribution of this to the strength of the hinge is ignored. The capacity of the Mesnager hinge is considerably less than that of a Freyssinet hinge.

4. Freyssinet hinges, in which little or no reinforcement passes through the throat and the concrete of the throat is designed to withstand compressive stresses considerably greater than its characteristic strength. Freyssinet hinges permit large rotations to take place at the throat whilst only small moments are transmitted to the structure.

These hinges should not be used where there is a risk of a collision with the structure causing damage or displacement of the hinge, nor, unless special precautions are taken to resist it, where there is a possibility of uplift under any loading condition.

4.3.11   OTHER TYPES OF BEARING

The above types of bearing may be used in combinations (e.g. a roller bearing surmounted by a rocker bearing, a sliding bearing with a pot or elastomeric bearing) to provide the necessary degree of freedom referred to in section 4.2. Other types of bearing may be used provided the performance requirements are met.

4.4   General design considerations

4.4.1   DESIGN LIFE

Bearings should normally be designed to last as long as the bridge in which they are situated. However, with some of the non-metallic materials in use today, although accelerated ageing tests would indicate a long life, insufficient experience of these materials in service conditions makes it impossible to be certain that bearings incorporating them will meet the serviceability requirements throughout the life of the bridge. Similarly, inadequate maintenance of metallic parts of bearings may reduce their serviceable life. It is, therefore, prudent for provision to be made for the inspection and, if found necessary, the removal and replacement of bridge bearings in whole or in part. Provision should be made for the installation of jacks necessary for the removal of bearings or any part thereof, insertion of shims or any other operation requiring lifting of the bridge deck from the bearings. Clamping plates, bolts or any other component that may require replacement during the service life of the bearing should be designed such that removal of the item from the bearing assembly within a moving structure can be achieved with minimal difficulty. Bearing plates should be attached to the structure in such a way as to facilitate the removal or replacement of the bearing. Adequate space should be provided around bearings to facilitate inspection and replacement and, where necessary, suitable handling attachments should be provided on the bearings. Means of access should not be overlooked. For bearings in halving joints access for inspection and replacement is practically impossible for slab type bridges and extremely difficult for beam and slab construction.

If there is a possibility of differential settlement, provisions should be made for jacking up the bridge deck and inserting metal shims or other suitable packing under the bearings unless, of course, the structure has been designed to accommodate the settlement or tilt in some other way.

4.4.2   DURABILITY

Bearings should be detailed without crevices and recesses that can trap moisture and dirt. The materials used in their manufacture and the method adopted for protection against corrosion should be such as to ensure that the bearings function properly throughout their life. It is important to ensure that dissimilar materials that can give rise to corrosive currents are not used together. Guidance on these aspects can be obtained from BS 5493 [6], PD 66484 [7,8]. Where the bearings are to be installed in particularly aggressive conditions, e.g. marine and corrosive industrial situations, special precautions will need to be taken.

4.4.3   MOVEMENT RESTRAINT

Where restraints are required to restrict the translational movement of a structure, either totally, partially or in a selected direction, they may be provided as part of or separate from the vertical load bearings. Restraint can be provided by separate dowels, keys or side restraints on sliding bearings. The relative merits of providing horizontal fixity by combined or separate horizontal and vertical load bearings depend upon the ratio of horizontal and vertical loads carried.

The load taken by dowels depends upon the shear and bending capacity of the dowel through the bedding and the bearing pressure on the concrete. There are several dowelling systems on the market today that have been developed and tested to take specific loads. Separate bearings to take horizontal loads may be necessary with large spans, but it may be more economic to specify bearings with an uprated vertical load to cater for horizontal loads on short spans.

In each case the restraints should allow freedom of movement in the desired direction(s). The forces generated by the restraints should be considered in the design of the structure. Where reliance is placed on friction to resist these forces, the lower bound value of friction coefficients obtained from available test data appropriate to the surface condition in service should be assumed. It is advisable to prevent such slip by means of additional measures such as bolting or by the use of bedding adhesive, rather than relying on friction alone.

Since bearing replacement may be required during the life of a structure, the provision of a restraint (e.g. dowels) through the bearings may cause difficulties, and alternative location of the restraints should be considered. In designing elastomeric bearings to withstand braking forces the effect of the braking movement on the deck expansion joint must be checked. For thick bearings, the effect of the total vertical deflection must also be considered.

4.4.4   UPLIFT

If the bearings will be subject to uplift, they and their fixings must be designed to limit separation of the parts. In particular, rubber should not be allowed to go into tension and sliding surfaces should not be allowed to separate. This would allow dust, grit and other abrasive or corrosive materials to enter and affect the sliding surfaces.

4.4.5   LIMIT STATES

The general statement is as given in section 3.7 for expansion joints. To meet the serviceability limit state for bearings the design should be such that they do not suffer damage that would affect their proper functioning or incur excessive maintenance during their working life. In the ultimate limit state the strength and stability of the bearings should be adequate to withstand the ultimate design loads and movements of the structure.

4.4.6   OUTER BEARING PLATES

Outer bearing or spreader plates are used to distribute the highly concentrated load at the bearing into the adjacent parts of a bridge structure and to ensure that the permissible stresses in concrete or other medium used in the abutments and deck are not exceeded. Various propositions have been put forward as to how these plates should be proportioned to achieve the desired object, some assuming a 45° spread of load, but BS 5400: Part 9 allows the effective area of distributed load to be taken as the contact area of the bearing plus the area within the uninterrupted dispersal lines drawn at 60° to the line of action of the bearing (Fig. 4.2). However, tests have shown that the load does not effectively spread to the corners of rectangular plates. For a detailed consideration of the relationship between distribution plate size and concrete stress see the paper by Buchler [9].

4.4.7 LOCATION OF BEARINGS

Bearings must be positioned such that they are able to operate as intended in the design of the structure. Any secondary effects resulting from either eccentric loading or movements not truly along a major axis of the bearing should be taken into account in the design of the bearing and related structural elements. The use of differing types of bearing in any one location is not recommended as it is difficult to cater fully for the interactive effects due to the different bearing characteristics. Care must be taken to ensure that dirt and debris cannot accumulate around bearings to impair their proper working, for example, by setting them on plinths with the top surface having a fall away from the bearing to remove any water that might collect. As discussed in section 4.4.1 they should be positioned such that they remain accessible for inspection and maintenance.

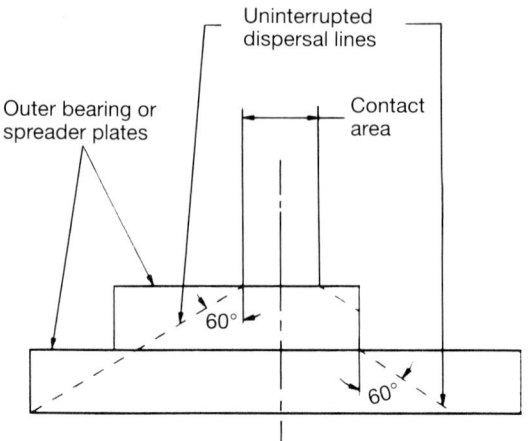

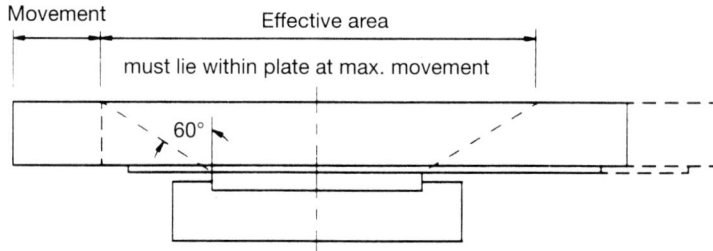

Fig. 4.2 Dispersal of load through plates.

4.5   Detail design considerations

4.5.1   ROLLER BEARINGS

Introduction

Recent developments in bridge design have resulted in bearings being required to take very high loads while at the same time accommodating large movements and some degree of rotation. Conventional mild steel bearings, if adapted to those conditions would require a nest of rollers to accommodate the vertical load and horizontal movement, surmounted by a rocker bearing (or in some cases a knuckle with large diameter pin) to accommodate the rotations. The problem has been overcome by developing special steels which allow high local bearing stresses.

Design of roller bearings

The ability of curved surfaces and plates to withstand deformation under load is dependent upon the hardness of the material of which they are made. The relationships between the local bearing or Hertz pressure, the hardness and ultimate tensile strength of steel for various types of bearing are shown in Fig. 4.3. It should be noted that there is not a constant relationship between hardness and yield stress but there is between hardness and ultimate strength. It must be emphasized that the hardness should not be just surface deep, which would tend to lead to cracking under load, but must penetrate well into the body of the metal. According to the Hertz theory [10, 11] for a cylindrical roller of radius R_1 on a concave surface of radius R_2, the maximum contact stress is given by

$$\sigma = 0.418 \left[PE \left(\frac{R_2 - R_1}{R_2 R_1} \right) \right]^{1/2}$$

where P is the load per unit length of roller.

To limit the deformation of the roller and plate to an acceptable level this contact stress is limited to 1.75 times the ultimate tensile strength σ_u

Thus

$$0.418 \left[PE \left(\frac{R_2 - R_1}{R_2 R_1} \right) \right]^{1/2} = 1.75 \, \sigma_u$$

whence

$$P = \frac{17.53}{E} \sigma_u^2 \left(\frac{R_2 R_1}{R_2 - R_1} \right)$$

In BS 5400: Part 9.1 the factor 17.53 is rounded to 18. For a roller of diameter D on a flat plate this becomes

$$P = \frac{8.76}{E} \sigma_u^2 D$$

or, as rounded up and quoted in BS 5400: Part 9.1,

$$P = \frac{18R}{E} \sigma_u^2$$

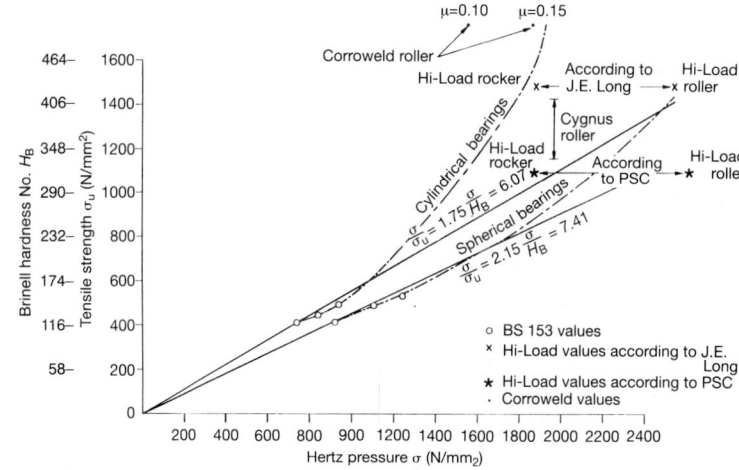

Fig. 4.3   Relationship between Hertz pressure, ultimate strength and hardness.

Similarly for a sphere of radius R_1 in a spherical seating of radius R_2 the maximum contact stress is given by

$$\sigma = 0.388 \left[PE^2 \left(\frac{R_2 - R_1}{R_2 R_1} \right)^2 \right]^{1/3}$$

Because of the point nature of the contact the stress is limited to 2.15 times the ultimate tensile stress σ_u.

Thus

$$0.388 \left[PE^2 \left(\frac{R_2 - R_1}{R_2 R_1} \right)^2 \right]^{1/3} = 2.15 \, \sigma_u$$

whence

$$P = \frac{170.15}{E} \sigma_u^3 \left(\frac{R_2 R_1}{R_2 - R_1} \right)^2$$

and for a sphere on a flat surface this reduces to

$$P = \frac{170.15}{E} R^2 \sigma_u^3$$

It is only necessary to consider the contact stresses in rollers and rockers at the serviceability limit state since the collapse condition is so obviously distanced from the serviceability limit. Local flattening of a roller or indentation of a plate, while increasing resistance to movement, would not prevent the bearing from supporting the structure. Nevertheless other parts of the bearing which can be in bending, shear or torsion should be assessed both for the serviceability and ultimate limit states in accordance with Part 3 of BS 5400 [12].

Types of roller bearing

The preference today is to use single roller bearings made of special high tensile alloy steels, such as the Hi-Load marketed by PSC Freyss-

Fig. 4.4 Glacier Cygnus roller bearings on M5 Huntworth Viaduct, Somerset. (Courtesy The Glacier Metal Co. Ltd.)

inet Limited or Glacier Metal Company's Cygnet bearings (Fig. 4.4), for carrying high loads. However, bearings containing multiple cylinders of lower quality steel or cast iron can be used. These require to be surmounted by a knuckle or rocker bearing to allow for rotation of the bridge deck. If more than two rollers are employed, the maximum permitted design loads given above for single rollers should be reduced by one-third to allow for uneven loading of the rollers caused by dimensional differences. To save space rollers can be flat sided Fig. 4.1(a). Such rollers should be symmetrical about the vertical plane passing through the centre and the width should not be less than one-third of the diameter or such that the bearing contact does not move outside the middle third of the rolling surfaces when the roller is at the extreme of its movement.

Some engineers prefer rollers having differing radii at the top and bottom as in Fig. 4.5. The distance moved is $r\theta + R\theta = h\theta$. This is independent of r and R provided $r + R = h$, the distance between the bearing plates. The advantage of this arrangement over a roller with a

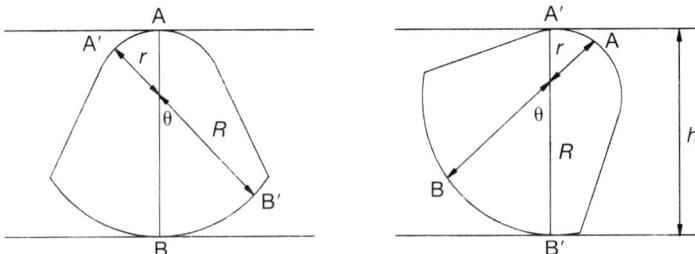

Fig. 4.5 Rollers with differing radii at bearing surfaces.

constant radius ($R = r$) is that the movement of the upper reaction point is reduced which is beneficial if the upper member is a steel girder with a bearing stiffener.

It is important that the bridge deck is stable under movement with non-cylindrical rollers (Fig. 4.6); thus

1. if $2r < h$ the bridge drops with movement away from the central position and is hence inherently unstable;
2. if $2r > h$ the bridge rises with movement away from the central position so the bridge is inherently stable.

Because of the geometry of this type of 'rocker' the two vertical reactions are not vertically in line. This produces a moment Pl which is resisted by a moment from the frictional force F of Fh. If l is too large, F will be so great that it can no longer be resisted by friction. This causes the roller to pop out (the orange pip effect; Fig. 4.7). The advantage of this type of bearing over a true roller is that, as the radius is larger for the same height, a greater load per unit length can be taken.

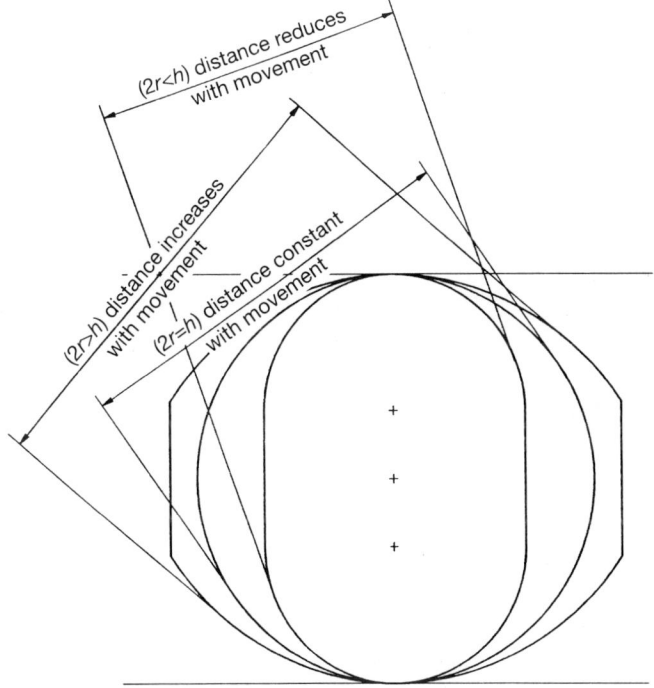

Fig. 4.6 Effect of geometry on stability.

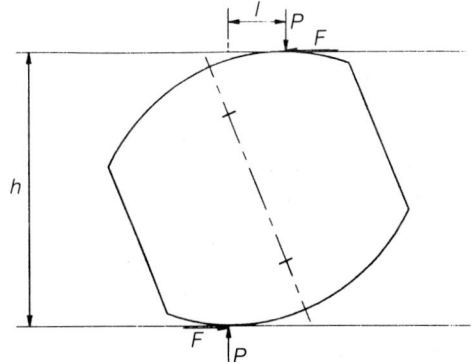

Fig. 4.7 Effect of non-coincident centres of surface radii on reactions.

Alignment

It is important that roller bearings are correctly aligned when installed and the alignment of the rollers is maintained during service and sliding of the roller on the bearing plates is prevented. In the case of Hi-Load bearings guidance is provided by flanged ends to the rollers which oversail and bear onto the ends of the bearing plates. If these flanges are called upon to carry high transverse loads, it is important to check the actual bearing stresses making allowances for any chamfers on the plates to accommodate the flange to roller fillets. Glacier and Kreutz roller bearings achieve the same result with a central flange running in grooves within the bearing plates. Sliding and twisting of rollers, which would cause increased friction by binding on the flanges, are usually prevented by some form of rack and pinion gear. The pitch circle diameter of the gear teeth must be the same as that of the rollers if the teeth are not to be damaged in service. A satisfactory basis of design for the gearing is to make the shear strength at the root of the tooth not less than the force required to make the roller slide.

Gearing is also beneficial in resisting the tendency of rollers in single roller bearings to squeeze out when the upper and lower plates are not parallel due to rotation of structural members, particularly when this type of bearing is used on long slender columns (Fig. 4.8). Normally this is prevented from happening by frictional forces, but could be a problem if roller friction is very low.

Friction of roller bearings

The allowable loads on roller bearings are usually sufficient to produce the first stages of plastic deformation leading to flats on the rollers and depressions in the bearing plates. The magnitude of the effect depends upon the hardness of the materials and the relationship of the contact stress to the yield stress of the material used. The value of the coefficient of rolling friction varies with the number of rolling repetitions over the same area, the initial value being some two to four times the value recorded after several cycles. In a bridge the normal daily travel will not occur regularly over the same piece of bed plate due to the seasonal variations in mean temperature and length of bridge. Thus the running-in period is going to last at least one year and the initial rolling

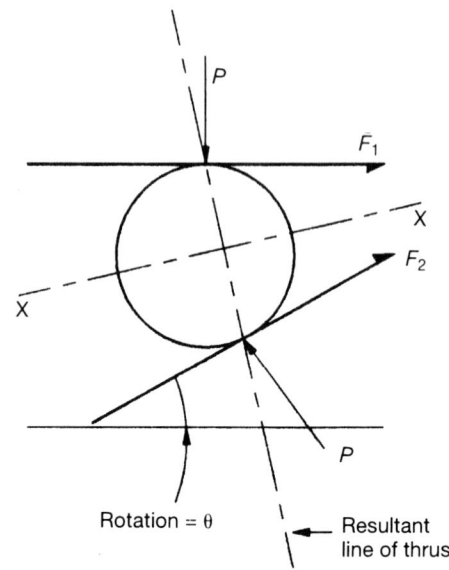

Fig. 4.8 Non-parallel bearing plates.

resistance will probably be the greatest. However, the steady state friction may be important if fatigue is being considered.

From tests carried out on high tensile steel rollers [13] the coefficient of initial rolling friction can approach 0.01. A factor of three is allowed for imperfect machining, poor alignment, dirt and corrosion giving the value of 0.03 quoted in BS 5400: Part 9.1 for normal steels. This is increased when more than two rollers are used in an assembly as it cannot be guaranteed that the loading will be equally distributed to all rollers. For rollers and bearing plates of specially hardened steel lower friction values can be used [1].

4.5.2 ROCKER BEARINGS

Rocker bearings, which are used at the fixed end of a bridge, or to provide rotation with multiple roller bearings, are designed using the

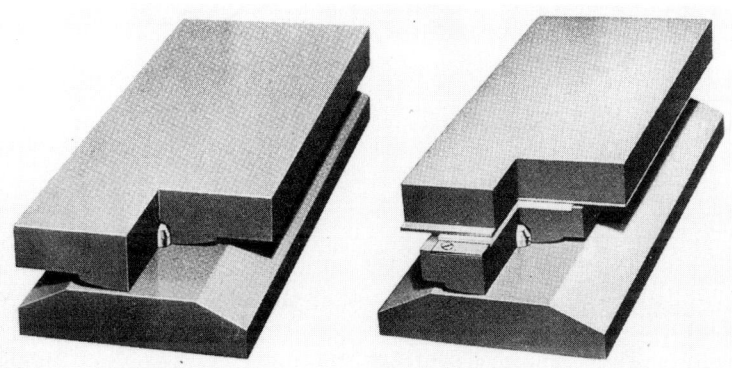

Fig. 4.9 BTR Solarbridge rocker bearing. Horizontal movement is facilitated by PTFE/stainless steel sliding surfaces. (Courtesy BTR Silvertown Ltd.)

same principles as for roller bearings and using the same materials. Relative horizontal movements of the parts must be prevented by dowel or other suitable action (Fig. 4.9).

4.5.3 KNUCKLE BEARINGS

Knuckle bearings (Fig. 4.10) provide only for rotation; other types such as roller or plane sliding bearings must be used in conjunction with the knuckle bearing to cater for translational movement. There are three main types of knuckle bearings; knuckle pin bearings, where the upper and lower curved surfaces slide around a central pin; leaf bearings, where the pin passes through a number of leaves attached alternatively to the upper and lower bearing plates, and curved sliding bearings. The last can have either cylindrical or spherical curved surfaces. Leaf bearings can be designed to resist uplift, if required.

Older forms of bearing were made of steel or cast iron. Friction could become quite high due to the effects of corrosion, as it is difficult to get at the sliding surfaces for maintenance. Grease grooves connected by drilled holes to end nipples can be used to overcome this

Fig. 4.10 Knuckle pin bearings under railway bridge over River Avon, Bath. (Courtesy R.G. Bristow.)

problem, but the modern practice is, generally, to concentrate on curved sliding surfaces, where the low friction properties of polytetrafluoroethylene (PTFE) can be employed.

As with roller bearings, the contact stresses need only to be considered at the serviceability limit. The design of knuckle bearings is based on the projected or plan contact area of upper and lower parts. In accordance with BS 5400: Part 9.1 the bearing pressure obtained by dividing the bearing load by the projected contact area (length of seating × diameter of pin) should not exceed

1. one half of the nominal yield stress of the weaker material used for the upper and lower part or 120 N/mm^2, whichever is the lesser, for all grades of steel;
2. 30 N/mm^2 for phosphor bronze;
3. 25 N/mm^2 for leaded bronze.

For permissible stresses on PTFE see section 4.5.5.

Other parts of the bearing are designed in accordance with the recommendations of BS 5400: Part 3. Lateral loads can be transmitted, if necessary, through flanges on the ends of the pin or directly through the leaves of leaf bearings.

4.5.4   LINK BEARINGS

Link bearings must be long enough between the pin end connections to limit their angular movements to not more than about ±2.9° and, thereby, avoid inducing horizontal forces greater than 5% of the vertical load. However, their overall length will also be dictated by the available construction depth. For a suspended span this would be limited to the girder depth and the compactness of the pin end details are important.

For a tension link, flat bar or high tensile screwed rods are appropriate (Fig. 4.11). Rod links with pins machined on the side of the solid end blocks provide a very compact end connection but maintenance requires careful consideration of the details. In compression the strut stiffness of an I section or a hollow section will be required.

The link pins are designed for shear and bearing but are also subjected to bending. This is governed by the fit of the pin in the pin hole,

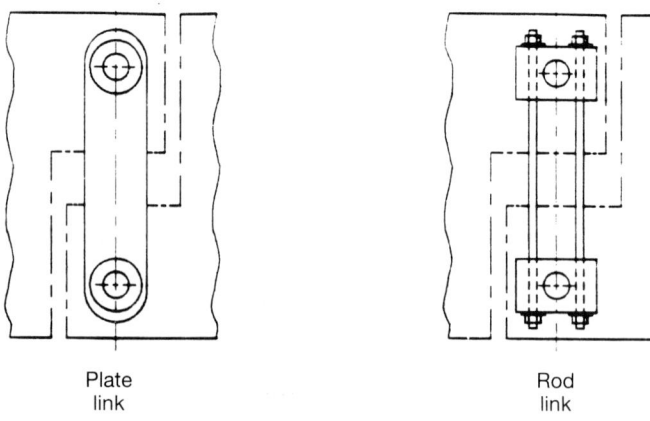

Plate
link

Rod
link

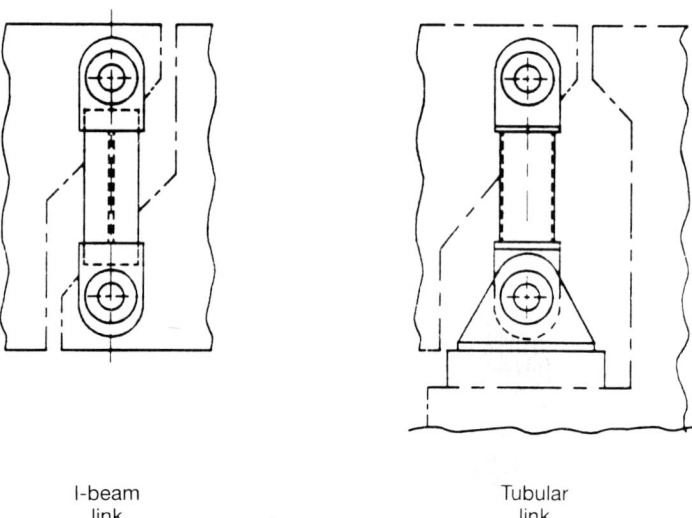

I-beam
link

Tubular
link

Fig. 4.11   Link bearing arrangements.

but for normal structural tolerances the assumption of uniformly distributed loading on contact surfaces will be valid. For compactness, pins are frequently made from higher tensile (heat treated) material selected from BS 970 [14]. Grease grooves and nipples are desirable to reduce friction and wear.

Design codes such as BS 5400 [12] give simple rules for proportioning the links at pin holes. These allow for the combined axial and bending stresses which exist at these locations. Since these effects are local it is usual to add reinforcing ring plates at the hole rather than increase the section of the link over its entire length.

Care must be taken to ensure that the chosen arrangement permits easy access to all surfaces for inspection and maintenance and consideration should also be given to ease of replaceability.

4.5.5 SLIDING BEARINGS

Introduction

Sliding bearings of a metal-upon-metal type have been in use for more than a century, generally bronze sliding against steel or cast iron with some sort of lubrication. These were very simple and cheap, but even when freshly and evenly lubricated the coefficient of friction is of the order of 20%. It was also found that, even with regular maintenance, the coefficient of friction increased to unacceptable limits. However, at a time when it appeared that sliding bearings were to become a historical curiosity, they were revitalized by the introduction of polytetrafluoroethylene (PTFE).

PTFE

PTFE is a fluorocarbon polymer, one of a group of plastics remarkable for their extreme chemical resistance, excellent dielectric properties and wide range of working temperature. They were first produced, other than on a laboratory scale, during the Second World War for use in the handling of highly corrosive products connected with the manufacture of uranium 235. However, the extremely low coefficient of friction was not utilized until some years after its development. The two features of chemical stability and low coefficient of friction make it an ideal material for bearings.

A variety of PTFE-based bearings are now available, the low-friction surfaces being used either to provide rotation by sliding over cylindrical or spherical surfaces, or to provide horizontal sliding movement, or a combination of both. The factor which is of greatest interest to design engineers is the coefficient of friction, this being a critical aspect in the design of structures incorporating PTFE sliding bearings.

PTFE sliding bearings (Figs 4.12–4.16)

Plane sliding bearings allow translation only, except for rotation about an axis perpendicular to the plane of sliding. As with multiple roller bearings rotation about a horizontal axis has to be accommodated by other means. The PTFE normally provides the lower sliding surface with the upper surface extending beyond the PTFE at the extremes of movements so that dust and dirt cannot settle on the lower PTFE surface to cause scouring and hence increase the friction. However, with this arrangement movement results in a shift in the point of application of load to the superstructure. In the case of a steel span this may result in an unacceptable eccentricity of loading under bearing stiffeners or diaphragms. The bearing can then be inverted provided that:

1. the extended (lower) sliding plate is kept clean by brushes attached to the upper part of the bearing or, preferably, enclosed by bellows. The disadvantage of the latter being that whilst being easily removable and replaceable for inspection and maintenance, they must remain sealed in service;

2. the substructure is designed for eccentric loading from the bearing.

The surface mating with the PTFE must be harder than the PTFE and corrosion resistant. Stainless steel is usually used for plane sliding bearings, but hard anodized aluminium alloy castings are used for curved sliding bearings. Curved sliding bearings are classed in BS 5400: Part 9 as knuckle bearings. They can be either cylindrical, which allows rotation about one axis only or spherical, which allows

a

Fig. 4.12 Examples of PTFE sliding bearings, (a) fluorocarbon slide bearings. (Courtesy Fluorocarbon Ltd); (b) BTR Solarbridge sliding bearings. (Courtesy BTR Silvertown Ltd); (c) BTR Solarbridge floating piston sliding bearing. (Courtesy BTR Silvertown Ltd.)

b

c

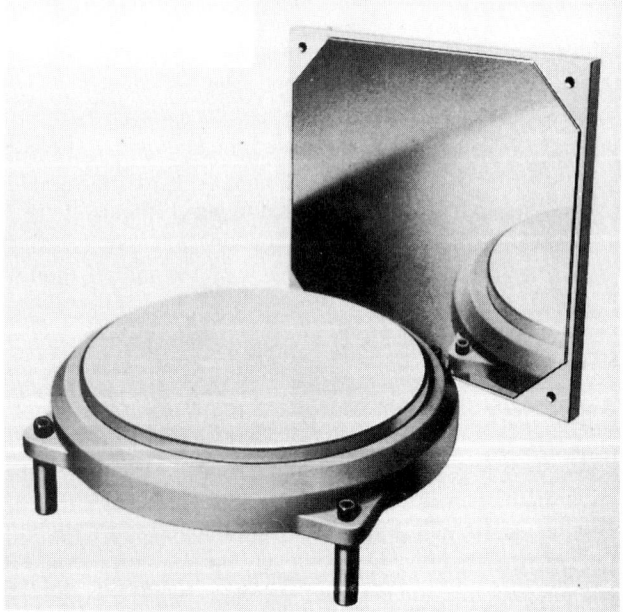

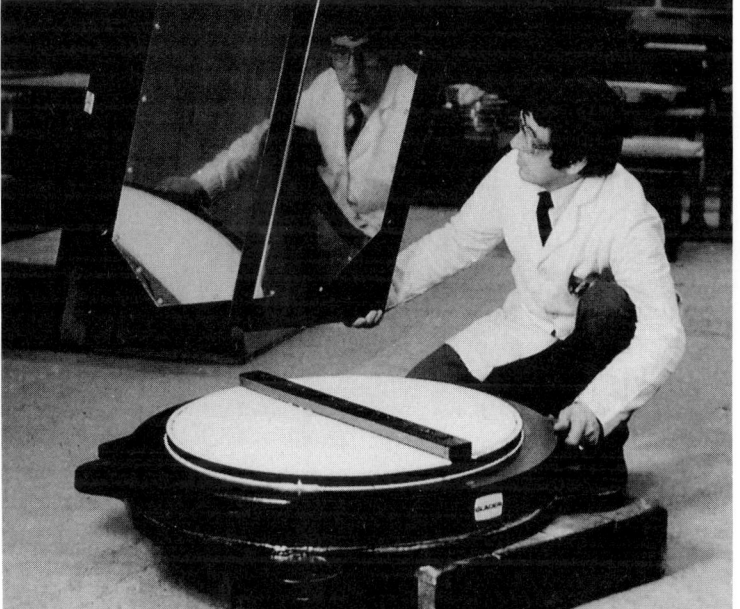

Fig. 4.13 Examples of guided sliding pot bearings, (a) CCL guided sliding pot bearing. (Courtesy CCL Systems Ltd); (b) base of Glacier 3000 t guided spherical bearing. (Courtesy The Glacier Metal Co. Ltd); (c) final inspection of Glacier Dualign K Series pot bearing. (Courtesy The Glacier Metal Co. Ltd.)

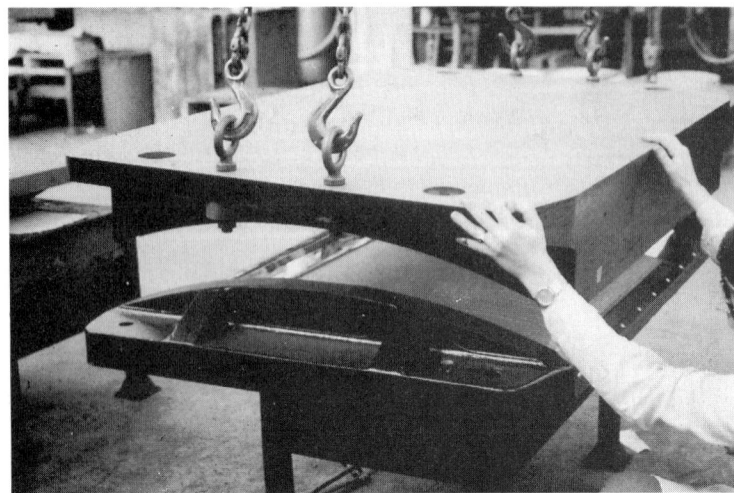

Fig. 4.14   Final assembly of Glacier 3400 t anti-clastic bearing. (Courtesy The Glacier Metal Co. Ltd.)

Fig. 4.15   Glacier spherical bearing with polished stainless steel convex surface. (Courtesy The Glacier Metal Co. Ltd.)

rotation about any axis. A special form of curved sliding bearing produced by the Glacier Metal Company and known as anti-clastic (Fig. 4.14) consists of two cylindrical curved sliding surfaces at right angles, allowing rotation about two orthogonal axes.

The PTFE can either be confined in a recess or be bonded to the backing medium. In either case it is essential that it is backed by a metal plate, the rigidity of which should be such that it retains its unloaded shape and resists shear forces under all loading conditions. It is usual for PTFE used in bearings to be provided with lubrication cavities, or dimples, as discussed later. The thickness of the PTFE is related to its maximum plan dimensions. Minimum thicknesses and maximum projections of recessed PTFE are laid down in BS 5400: Part 9 (Tables 4.1 and 4.2).

The thickness of the mating stainless steel sheet is related to the difference in dimension between the PTFE and stainless steel in the direction of travel. The maximum recommended thicknesses are given in BS 5400: Part 9 (Table 4.3).

It is essential that the stainless steel remains flat throughout its service life and corrosion is prevented by keeping moisture from getting between the stainless steel and its backing medium. This can be done by attaching the stainless steel to its backing plate by continuous welding along its edges. Alternatively the stainless steel can be fixed to its backing medium with fasteners supplemented by either peripheral sealing or bonding over the full area of the stainless steel sheet. The

Table 4.1   Dimensions of confined PTFE

Maximum dimension of PTFE (diameter or diagonal) (mm)		Minimum thickness (mm)	Maximum projection above recess (mm)
≤ 600		4.5	2.0
> 600	≤ 1200	5.0	2.5
> 1200	≤ 1500	6.0	3.0

Fig. 4.16 Mageba sliding bearings before assembly. (Courtesy Mageba S.A.)

Table 4.2   Thickness of bonded PTFE

Maximum dimension of PTFE (diameter or diagonal) (mm)	Minimum thickness (mm)
⩽ 600	1.0
> 600 ⩽ 1200 (max)	1.5

Table 4.3   Thickness of stainless steel sheet

Dimensional difference between PTFE and stainless steel [1] (mm)	Minimum thickness of stainless steel (mm)
⩽ 300	1.5
> 300 ⩽ 500	2.0
> 500 ⩽ 1500	3.0

[1] A dimensional difference in excess of 1500 mm requires special consideration.

Table 4.4   Allowable sliding bearing pressures for pure PTFE

Design load effects	Maximum average contact pressure (N/mm^2)		Maximum extreme fibre pressure (N/mm^2)	
	Bonded PTFE	Confined PTFE	Bonded PTFE	Confined PTFE
Permanent design load effects	20	30	25	37.5
All design load effects	30	45	37.5	55

method of attachment should be capable of resisting the full frictional force set up in the bearing at the serviceability limit state.

The bearing surfaces need be designed only for the serviceability limit state, but the remainder of the bearing should be designed to satisfy the requirements for both serviceability and ultimate limit states. Only pure PTFE is used for normal sliding bearings, but filled PTFE is permitted to take higher stresses on guides. The maximum permitted contact pressures on pure PTFE are given in BS 5400: Part 9 (Table 4.4).

For calculations of pressures, the contact surfaces may be taken as the gross area of the PTFE without deduction for the area occupied by lubrication cavities. In the case of curved surfaces, the gross area should be taken as the projected area of the contact surface.

Horizontal forces applied to curved sliding surfaces tend to separate the contact surfaces of bearings. This can give rise to chatter, uneven wear and ingress of dirt between the sliding surfaces leading to corrosion and increased friction. Therefore, a check should be made to ensure that this tendency is adequately resisted by the coincident vertical loads. The calculations for the destabilizing horizontal force and restoring vertical force should be based on the recommendations regards overturning given in BS 5400: Part 2 [15].

Wear of PTFE

PTFE wear is dependent upon
1. pressure;
2. temperature range;
3. rubbing speed;
4. length of travel.

There is considerable information on the effects of the first two but little data on the actual movement of bridges, particularly under traffic loads. In most European countries the working pressures are specified with the respective coefficients of friction.

Apart from some individual measurements, especially on railway bridges, few data are yet available on actual sliding speeds that occur in practice, but it is possible mathematically to determine

Fig. 4.17 Permanent bearing (Glacier anti-clastic bearing) used for slide-in with stainless steel sheet on pier top. (Courtesy The Glacier Metal Co. Ltd.)

sliding speeds in bearings. This is easier for railway bridges than road bridges. For the standard endurance tests on PTFE carried out in Germany, sliding speeds of 0.4 mm/s are used for low temperatures and 2.0 mm/s for room temperatures. In bridges, temperature, creep and shrinkage produce negligible sliding speeds but the effects of the large number of short sliding movements from traffic are more critical on bearing life. Figure 4.17 shows PTFE sliding bearings (Glacier anti-clastic) installed on a railway bridge.

The determination of the total sliding movement in road bridges is subject to the same difficulties as the determination of sliding speeds and there is a need for measurements to be made on a variety of bridge structures to establish parameters for design and maintenance of bearings. Kauschke and Baigent [16] state that PTFE bearings should be satisfactory for up to 10 km and possibly 20 km of sliding movement and gives an example to show a bearing life of 32 years assuming a

total movement limit of 5000 m. They make the point that PTFE surfaces of sliding bearings must be considered as wearing parts and be replaceable. For 'live' structures (section 1.5) special arrangements need to be made to monitor the wear of working parts of bearings and for PTFE bearings the projection of the PTFE above its backing plate or recess should be regularly monitored.

The operational life of a sliding surface is also related to the maintenance of a low coefficient of friction. Attempts have been made to pump new grease into the sliding surfaces. However, grease hardens over the years making its replacement difficult. Therefore, replacement of the PTFE sheets is a more practical solution.

Friction with PTFE surfaces

Tests [17–20] have indicated that the coefficient of friction of stainless steel plate sliding on PTFE varies with the speed of movement, contact stress between plate and PTFE, temperature, finish of mating surface and previous loading history. In general the coefficient of friction between PTFE and mating surface reduces with

1. reducing speed of movement;
2. increasing compressive stress;
3. increasing temperature;
4. increasing smoothness of mating surface.

The tests also show the initial static coefficient of friction, which relates to the force required to start movement, to be higher than the dynamic coefficient of friction which relates to the force to maintain movement, although the difference in magnitude between the two tends to reduce with increasing cycles of movement. This is illustrated in Fig. 4.18. During the early cycles, transfer of PTFE onto the mating surface takes place leading to a reduction in friction, but after continuous running this tends to wear off giving rise to an increase in friction until a state of equilibrium is reached where loss of PTFE due to abrasion balances the PTFE being transferred. Lubrication can be used to reduce the initial friction, but in time the lubricant will be squeezed out and the coefficient of friction will tend to return to the dry run-in value. Therefore, for the purposes of design the long-term value of the

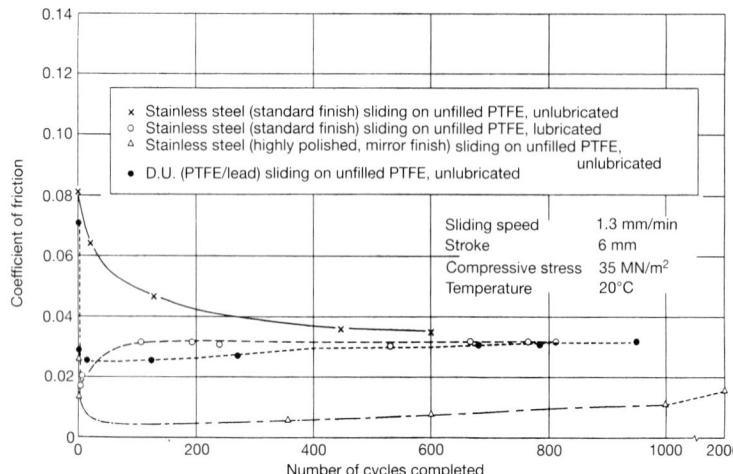

Fig. 4.18 Variation of friction with type of mating surface. (Based on TRRL Report No. LR 491.)

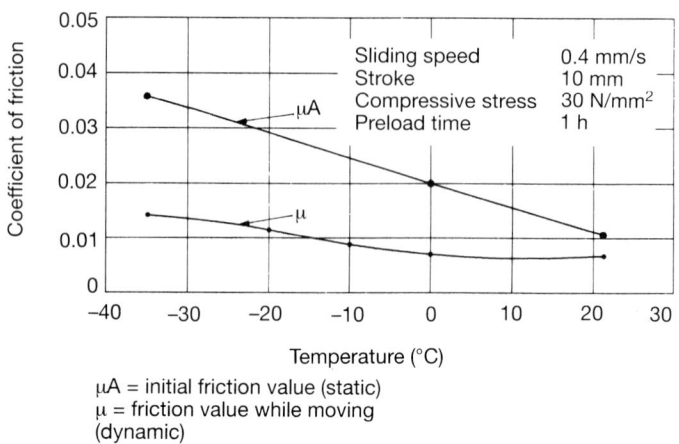

μA = initial friction value (static)
μ = friction value while moving (dynamic)

Fig. 4.19 Relationship between coefficient of friction and temperature of PTFE.

coefficient of friction is used. Silicone greases have been found especially suitable for lubrication, having proved effective with temperatures below –35°C. They do not resinize nor attack the material of the sliding surfaces. Tests on PTFE with lubrication pockets sliding on stainless steel over 5000 m of slide path show insignificant wear and virtually no increase in coefficient of friction [16].

The effect of temperature of the sliding surface and increasing compressive stress on the coefficient of friction is shown in Figs 4.19–4.21. The effect of temperature and the influence of surface finish are illustrated in Fig. 4.22 and the relationship between coefficient of friction and sliding speed is shown in Fig. 4.23.

When dealing with structural movements at low temperatures one must bear in mind that ice may form on the sliding surfaces, although the bond with the PTFE is likely to be poor, and also part of the PTFE in the bearing will be moving on a little used part of the steel slider.

For stainless steel sliding on pure PTFE which is continuously lubricated, Table 4.5 gives values for design purposes from BS 5400: Part 9.

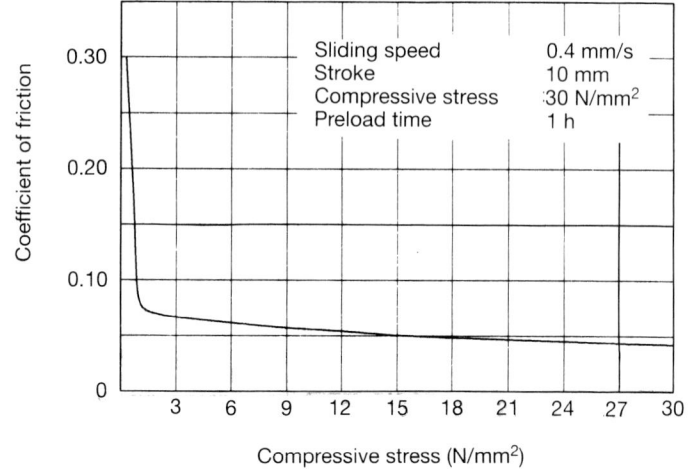

Fig. 4.20 Relationship between coefficient of friction and bearing pressure of PTFE.

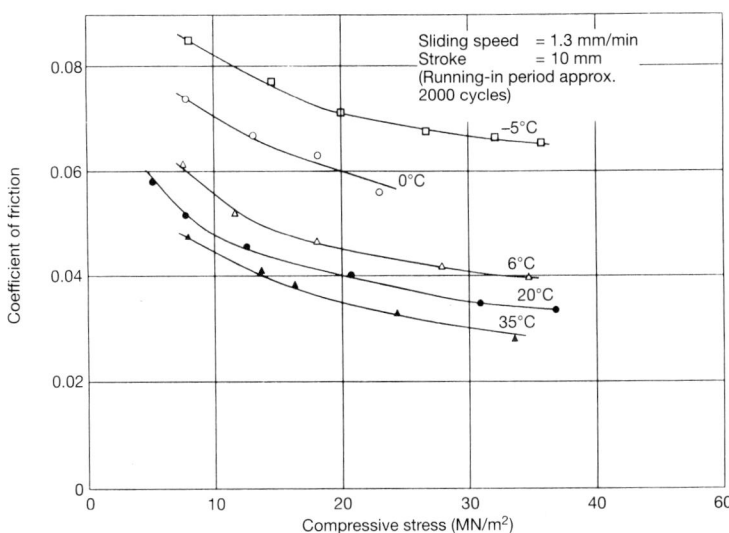

Fig. 4.21 Influence of sliding plate temperature and pressure on the peak coefficient of friction of unfilled PTFE. (Based on TRRL Report No. LR 491.)

Table 4.5 Design friction coefficients for PTFE on stainless steel

Bearing stress (N/mm²)	Coefficient of friction
5	0.08
10	0.06
20	0.04
30 and over	0.03

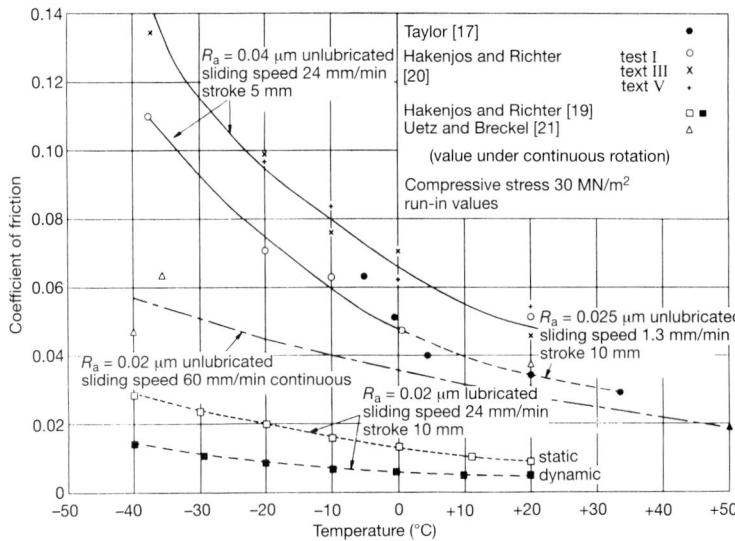

Fig. 4.22 Effect of temperature and surface finish on coefficient of friction of unfilled PTFE.

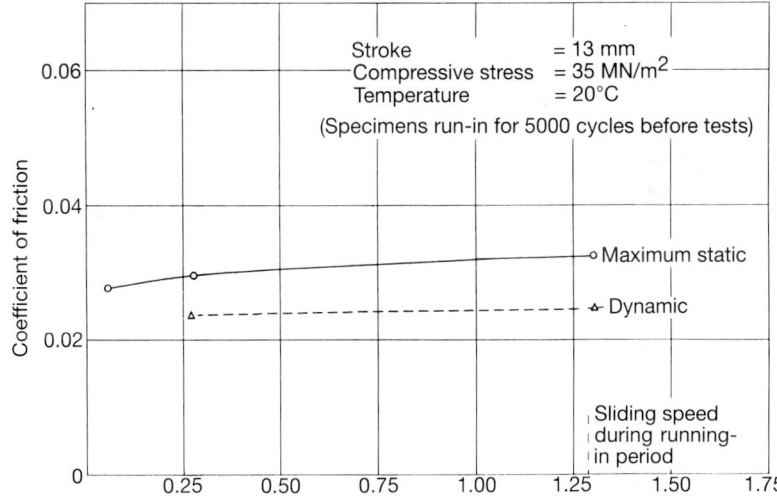

Fig. 4.23 Relationship of sliding speed to coefficient of friction of unfilled PTFE. (Based on TRRL Report No. LR 491.)

These values can be used for temperatures down to −24°C. For un-lubricated PTFE on stainless steel, BS 5400: Part 9 recommends twice these values should be used, and for filled PTFE four times these values are recommended.

Plain PTFE should be used if extremely low friction values are required. Compared with plain PTFE, filled PTFE has the advantage of a higher pressure resistance and increased wear resistance. Glass, carbon, graphite and bronze powder have proved to be suitable fillers.

With a large number of spans in a continuous structure, the statistical probability of all the bearings carrying a maximum friction load is reduced and it would be reasonable to design for an overall value somewhat lower than the maximum. This is relevant when one considers friction on all the bearings combined with, say, wind and centrifugal forces. A combination of forces might take account of the maximum centrifugal effect developed from vehicles travelling at full speed over a given length of structure, and vibrations from fast-moving traffic which would tend to promote free slip action; it would also allow for the fact that traffic is unlikely to be travelling fast when wind speeds are high or the extremes of temperature are being experienced, and that maximum wind pressure is unlikely to affect the whole length of a long structure at a given instant of time.

Under these effects, an anchorage may be designed to withstand the sum of the 5% bearing friction excluding wind and centrifugal force. If an anchorage is located approximately in the centre of a long structure, the difference in friction occurring between the two opposing lengths can be taken into account. One might take 7% from one half, counteracted by 3% from the other half. When looking at the forces in friction in combination with the maximum effects of wind and centrifugal force, 3% has been taken and it may even be possible to reduce it to 1%. It is important to relate the friction value being taken to precise loadings occurring in combination with other forces as it cannot be stressed too often that the value of the coefficient of friction changes with contact stress and other factors.

When large horizontal forces are being accommodated, one has to consider whether provision of PTFE rubbing surfaces are necessary to reduce the forces generated. It can be seen that, with a vertical sliding

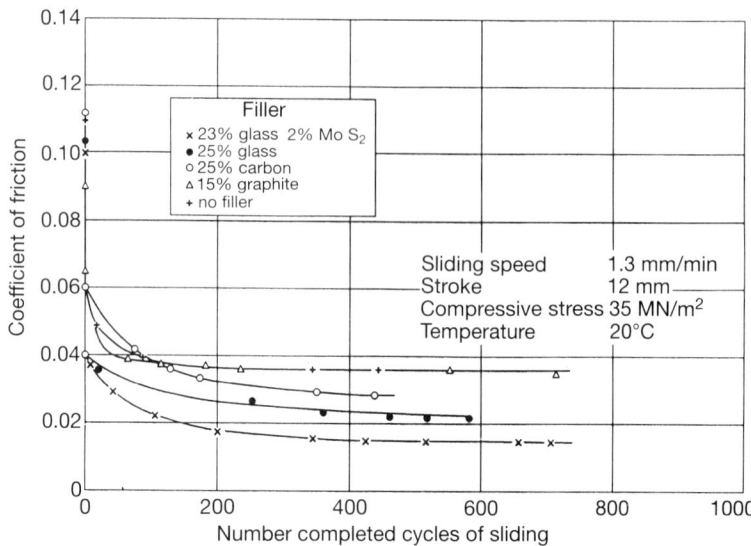

Fig. 4.24 Coefficient of friction of filled PTFE materials. (Based on TRRL Report No. LR 491.)

friction value of 5% and a steel-to-steel rubbing contact for lateral loading of say 25% to 75% of the vertical loading, an incompatible loading situation on the substructure can arise.

In these circumstances, to withstand higher stresses and reduce wear PTFE with glass fibre or other filling is often used. The effect of various fillings in PTFE on the coefficient of friction is shown in Fig. 4.24. This would seem to indicate that the use of fillers lowers the coefficient of friction. This is not borne out by other tests [21], and the generally accepted view is that fillers increase the coefficient of friction as referred to above.

Setting tolerances for PTFE bearings

One aspect that requires attention is the tolerance permitted in the setting of sliding bearings. Recommended tolerances are given in clause 8.3.2 of Section 9.2 of BS 5400: Part 9 [1].

Suppose a sliding bearing is to be set truly horizontal but is tilted in the direction of sliding by 10 mm in 1 m (1/8 in. in 1 ft). This inaccuracy is responsible for generating an additional apparent friction force of about 1% when the structure slides uphill – a high proportion of the true sliding friction for which proper allowance has therefore to be made.

A similar effect becomes apparent when one considers a change in rotation of the bridge deck relative to the supporting structure. Rotations may arise from any loading producing flexure such as prestressing forces or flexibility of the substructure.

Preferred positioning of curved sliding surfaces

When PTFE is employed on curved surfaces, attention should be paid to the optimum geometrical arrangement to reduce movement and therefore wear on the sliding surfaces. Fig. 4.25 illustrates such a consideration.

4.5.6 ELASTOMERIC BEARINGS

Introduction

In the past some form of roller bearing or sliding plate bearing has been used to accommodate comparatively small movements, say 6–125 mm (1/4–5 in), with or without rotation. Both these types of bearing are not really cheap in first cost and in addition require long-term maintenance, which they tend not to receive.

In recent years, bridge designers have turned to elastomeric pad bearings (Figs 4.26 and 4.27). These have advantages of easy installation, no maintenance and low cost. Elastomeric pads also have the advantage of not being susceptible to seizure. For instance, in roller bearings, because of the small movements which normally occur during the service of a bridge, there is a tendency to form flats over a small arc of the roller and, if lubrication is neglected, this may eventually cause seizure. Steel bearings can also give problems in dusty or corrosive conditions. Current experience indicates that, when they are properly designed and manufactured, elastomeric pad bearings have a life expectancy comparable with that of the bridge. Lindley [22] refers

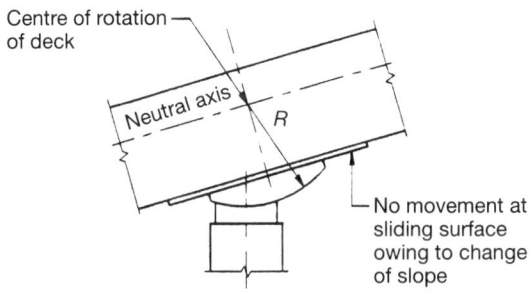

(a)

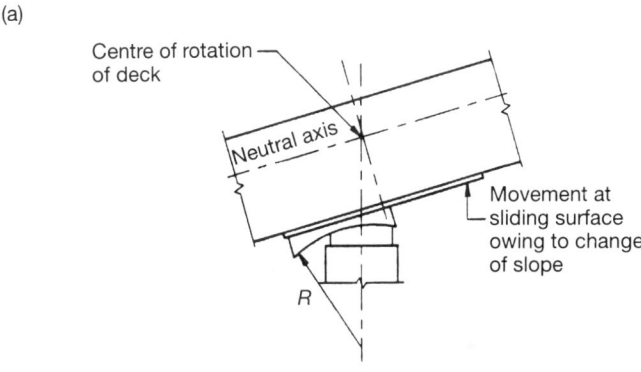

(b)

Fig. 4.25 Effect of rotation related to deck centroid where centre of curved surface is (a) coincident; (b) not coincident with centre of rotation of deck.

to the use of plain rubber pads supporting a viaduct superstructure in Australia, installed in 1889 and still in service in 1981 with degradation of the rubber at the surface limited to about 1 mm.

Rubber bearings accommodate horizontal movements by shearing and have enough shearing flexibility to avoid transmitting high loads to the bridge supports. Horizontal movements in more than one

Fig. 4.26 Section through elastomeric bearing. (Courtesy Victualic Industrial Polymers.)

direction can be accommodated. The vertical stiffness is such that significant changes in height under load are avoided.

With experience, it has been found that rubber bearings are extremely versatile and can be used for many an unexpected application by a suitable arrangement (Fig. 4.28). For example, bearings with a lead plug for damping movement under seismic forces have been suggested [23].

Materials

Some confusion has developed over the relative merits of natural and synthetic rubbers, probably attributable to commercial interests and a

Fig. 4.27 Elastomeric bridge bearing under test. (Courtesy Victualic Industrial Polymers.)

lack of specialized knowledge by civil engineers. For instance, natural rubber was originally excluded from the AASHO specification in favour of neoprene, but it has been included since the beginning of 1968. The most common material used in Britain is vulcanized natural rubber. Synthetic rubbers such as neoprene and chlorobutyl are used. They have the advantage of lower basic material cost, but, compared with natural rubber, the ratio of elastic modulus to shear modulus is poor. Also neoprene has a poorer performance and tends to become brittle at low temperatures. Chlorobutyls have good abrasion resistance and are sometimes bonded to natural rubber fenders in dock work.

The choice of specifying natural or synthetic rubbers depends on the conditions under which the bearings will operate. For instance, where temperatures are likely to drop to –10°C, natural rubber is preferable. Chloroprene rubber is superior where oil or grease contamination is likely to be encountered and also where the temperature might exceed 60°C. Nitrile is used where especially corrosive crude oil is likely to be in contact with the pads.

Some special properties required of the rubber for use in bridge bearings are that it must have a good resistance to the action of oils, weather, atmospheric ozone and the extreme temperatures to which the bearing is subjected.

Various compounding ingredients are incorporated in a natural rubber compound before it is vulcanized. Some of these ingredients are necessary for the vulcanization process, while others assist or accelerate it; chemicals such as antioxidants and antiozonants protect the rubber; reinforcing fillers, notably carbon black of which there are many varieties, stiffen the rubber, whereas oils soften it. Normally oil is only used as a processing aid to facilitate mixing of these ingredients with the raw rubber. (Although the word rubber is applied to the raw material, it more generally refers to the vulcanized material or vulcanizate. Vulcanized rubber is also referred to as elastomer.)

The vulcanization (or curing) of the compounded rubber is usually carried out under pressure in metal moulds at a temperature of about 140°C and takes from a few minutes to several hours depending on the type of vulcanizing system being used and the size of the component. The finished component has the shape of the mould cavity.

Design of elastomeric bearings

Rubber, although elastic, is almost incompressible, its modulus of bulk compression being similar to that of water. In order to function as a compression spring, therefore, the sides must bulge when a block of rubber is subjected to a vertical load. By limiting its freedom to bulge, the vertical deflection can be reduced and its stiffness increased. This is usually achieved by bonding reinforcing plates of steel between layers of rubber, the whole being surrounded by a thin layer of rubber to protect the metal plates from corrosion and to take up any irregularities in the supporting medium.

Some elastomeric bearings are made with vertical holes. These holes pass through the rubber layers and reinforcing plates. Dowels are placed in these holes to control the alignment of the reinforcement during manufacture and for locating the bearings in a structure. Since the reinforcement is a tension element the holes can affect the behaviour of the bearings. Tests [24] have shown that the stiffness of the bearing is reduced under service load and the holes have the effect of reducing the average stress level at which yielding of the reinforcement takes place as well as reducing the ultimate load capacity. In these circumstances it is suggested that plates with holes should be thicker than those without.

Tolerance of layer thickness is also critical as tests [24] indicate that poor tolerance control causes greater compression strains and deflections and can cause early initiation of yield.

The stiffness of a rubber bearing in compression, when the loaded surfaces are prevented from slipping (by bonding or otherwise) depends upon the shape factor S which is defined as the ratio of one loaded area to the total force-free surface area.

For plain pad bearings,

$$S = \frac{A}{l_p t_e}$$

where

A is the overall plan area of the bearing;

Fig. 4.28 Elastomeric bearings used for supporting prestressed concrete pressure vessels in nuclear power stations. (Courtesy CCL Systems Ltd), (a) Torness Nuclear Power Station; (b) Heysham Nuclear Power Station.

l_p is the force-free perimeter of the bearing, including that of any holes if these are not later effectively plugged;

t_e is the effective thickness of elastomer in compression, which is taken a $1.8t$, where t is the actual thickness of elastomer.

Note that for a rectangular bearing without holes,

$$l_p = 2(l+b)$$

where l is the overall length of the bearing and b is the overall width of the bearing.

For laminated bearings, the shape factor S for each individual elastomer layer is given by the expression

$$S = \frac{A_e}{l_p t_e}$$

where

A_e is the effective plan are of the bearing, i.e. the plan area common to elastomer and steel plate, excluding the area of any holes if these are not later effectively plugged;

l_p is as defined above;

t_e is the effective thickness of an individual elastomer lamination in compression; it is taken as the actual thickness, t_i, for inner layers, and $1.4t_i$ for outer layers.

Note that for a rectangular bearing without holes,

$$A_e = l_e b_e \text{ and } l_p = 2\,(l_e + b_e)$$

where l_e is the effective length of the bearing (equals length of reinforcing plates) and b_e is the effective width of the bearing (equals width of reinforcing plates).

The shear stiffness of elastomeric bearings is unaffected by the inclusion of steel reinforcing plates and the shear strain is usually limited to prevent fatigue problems. Under BS 5400: Part 9.1 the shear strain ξ_q due to translational movement given by the expression

$$\xi_q = \delta_r / t_q$$

should not exceed 0.7, where

δ_r is the maximum resultant horizontal relative displacement of parts of the bearing obtained by vectorial addition of δ_b and δ_l;

δ_b is the maximum horizontal relative displacement of parts of the bearing in the direction of dimension b of the bearing due to all design load effects (Fig. 4.29);

δ_l is the maximum horizontal relative displacement of parts of the bearing in the direction of dimension l of the bearing due to all design load effects (Fig. 4.29);

t_q is the total thickness of the elastomer in shear.

A similar value is used in the UIC Code [25].

BS 5400: Part 9.1 limits the mean design pressure (V/A) on elastomeric bearings to GS or $5G$, whichever is the lesser, where G is the shear modulus of the elastomer and S is the shape factor. The UIC Code allows $2GS$. Values of shear modulus G and bulk modulus E_b for natural rubber of varying hardness are given in Table 4.6.

At subzero temperatures crystallization causes stiffening of elastomers: the process is reversible. For bearings used at low temperatures the variation in G should be established by testing. However

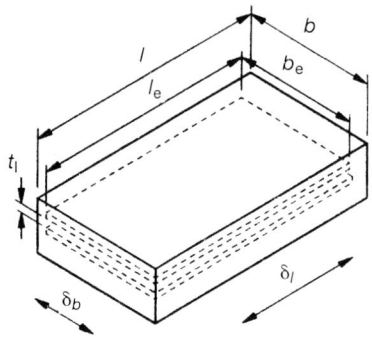

Fig. 4.29   Diagrammatic representation of elastomeric bearing.

Table 4.6   Elastic constants

Hardness (IRHD (± 2))	Shear modulus G (N/mm²)	(lb/in²)	Bulk modulus E_b (N/mm²)	(lb/in²)
35	0.38	55	2000	290 000
40	0.45	65	2000	290 000
45	0.53	77	2030	294 000
50	0.63	91	2060	299 000
55	0.75	109	2090	303 000
60	0.89	129	2120	307 000
65	1.04	151	2150	312 000
70	1.22	177	2180	316 000
75	1.42	206	2210	320 000

Table 4.7   Dynamic stiffening factors

	Hardness (IRHD)			
	40	50	60	70
Shear modulus G (N/mm²)	0.45	0.63	0.89	1.22
Dynamic stiffening factor D	1.1	1.25	1.5	1.9

BS 5400: Part 9.1 suggests that, in the absence of test data, values of G may be obtained by multiplying the values in Table 4.6 by $1-T/25$ where $T(°C)$ is the minimum shade air temperature.

Because natural rubber is a visco-elastic material the shear modulus is also dependent on the frequency or rate of application and the strain amplitude. Therefore for rapid strain variations, as for example due to live loads in bearings under railway bridges, G should be multiplied by the dynamic stiffening factors given in Table 4.7. The effect of temperature and frequency in the dynamic shear modulus is illustrated in Fig. 4.30.

Even under uniform compression strain, the compression stresses and the shear strains in an elastomer slab are markedly non-uniform

and depend on the shape factor S. The maximum shear strain occurs at the edge of bonded plates and its value ξ_b due to compression strain is given by the equation

$$\xi_b = 6S\xi_c$$

In the above equation ξ_c should not contain the strain due to bulk compression of the elastomer, as it does not introduce shear. When a slab is subjected to shear, the vertical reaction V will act on a reduced effective plan area

$$A_1 = A_e \left(1 - \frac{\delta_b}{b_e} - \frac{\delta_1}{l_e}\right)$$

approximately. This should be taken into account when calculating the values of ξ_c to be used but not in the calculation of S.

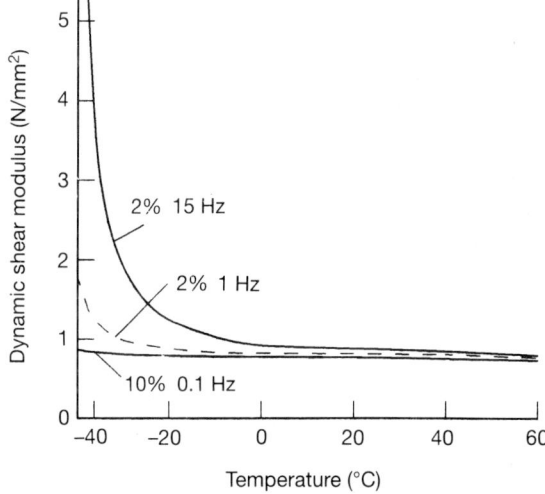

Fig. 4.30   Effect of temperature and frequency on the dynamic shear modulus of natural rubber of about 50 IRHD (percentage strain amplitudes shown).

In order to provide an adequate margin against the effects of fatigue, the sum of the nominal strains due to all load effects, ξ_t as given by the expression

$$\xi_t = k \, (\xi_c + \xi_q + \xi_\alpha)$$

should not exceed 5, where

  k is a factor equal to 1.5 for live load effects; 1.0 for all other effects (including wind and temperature);

  ξ_c is the nominal strain due to compressive loads, where ξ_c is given by the expression

$$\xi_c = 1.5 V / G A_1 S$$

  ξ_q is the shear strain due to translational movements;
  ξ_α is the nominal strain due to angular rotation, where ξ_α is given by the expression

$$\xi_\alpha = (b_e^2 \, \alpha_b + l_e^2 \, \alpha_l) / 2 t_i \sum t_i$$

  V is the design vertical load effect;
  A_1 is the reduced effective plan area due to the loading effects, where A_1 is given by the expression

$$A_1 = A_e \left(1 - \frac{\delta_b}{b_e} - \frac{\delta_l}{l_e} \right)$$

  α_b is the angle of rotation across the width, b, of the bearing (in radians);
  α_l is the angle of rotation (if any) across the length, l, of the bearing (in radians);
  t_i is the thickness of the individual layer of elastomer being checked;
  $\sum t_i$ is the total thickness of elastomer in the bearing. Other symbols are as defined above.

The limit $\xi_t = 5$ is an empirical value which has been found from fatigue tests on three types of elastomeric bearing to fit best the limit-ing criterion for a strain calculated by the method given here. It should not be taken to reflect the ultimate strain of the material. The tests were carried out by ORE, the Office for Research and Tests (Experience) in connection with the preparation of the UIC Code for the use of rubber bearings for rail bridges [25].

Rotation α due to angular deflection of a bridge deck under load will produce a tilt between the faces of a bearing, causing tensile strains in elastomers. In bonded elastomeric bearings such strains can lead to internal rupture and the ingress of ozone, and it is important, therefore, that net tensile strain should be avoided at the edges of this type of bearing, i.e. $\xi_t > 0$.

It is not possible, in the light of present knowledge, to specify a permissible amount of tilt appropriate for all conditions likely to arise (i.e., combination of compression, shear and rotation). As a general indication, elastomeric bearings designed so that the deflection at the edge of the bearing due to critical rotation does not exceed the deflection Δ under vertical load should normally be satisfactory; thus

$$\frac{\alpha_b b_e}{2} + \frac{\alpha_l l_e}{2} < \Delta$$

BS 5400: Part 9 appears to have applied a factor of safety of 1.5 to this expression. For practical purposes elastomeric bearings supporting precast concrete beams or steel girders should be designed to withstand a minimum rotation of 0.01 radians. In portals and arches elastomeric bearings can be used at the point of rotation.

When the thickness of an elastomer slab exceeds four times its least plan dimension there may be a significant reduction in the horizontal stiffness. To avoid the possibility of instability the total elastomer thickness should not exceed one quarter of the least lateral dimension or, in the case of laminated bearings,

$$\frac{V}{A_1} < \frac{2 b_e G S^1}{3 \sum t_i}$$

where S^1 is the shape factor for the thickest layer of elastomer and $\sum t_i$ is the total thickness of all layers.

Considering that the rubber behaves like a fluid, the application of a vertical load V, to an effective plan area A_1, produces a stress V/A_1, in all directions. In a rubber layer of thickness t_i, the total horizontal force produced per unit length is therefore $t_i V/A_1$, half of which is resisted by the steel plate above and half by the plate below. Thus the thickness h of a steel plate without holes having a working tensile stress σ_s between rubber layers of thickness t_1 and t_2, follows from $h = (t_1 + t_2)V/2A_1\,\sigma_s$.

However, tests indicate that the stress is not uniform across the plate, being a minimum at the edges with a maximum occurring at the centre of the plate, where the local value is about 1.5 times the average. It is therefore necessary to modify the above formula to

$$h = \frac{3(t_1 + t_2)V}{4A_1\sigma_s}$$

For the ultimate limit state it is necessary to relate this to the yield stress of the steel plates. Therefore it is necessary to introduce a safety factor of 1.75 to provide the expression for the minimum thickness of steel reinforcing plates given in BS 5400: Part 9.1:

$$\frac{1.3V(t_1 + t_2)}{A_1\sigma_s} \geq 2 \text{ mm}$$

where

V and A_1 are as defined above;

t_1 and t_2 are the thicknesses of elastomer on either side of the plate;

σ_s is the stress in the steel, which should be taken as not greater than the yield stress, nor greater than 120 N/mm^2 for plates with holes and 290 N/mm^2 for plates without holes.

The deflection of a single elastomeric pad under a vertical load V is given by the expression

$$\delta = \frac{Vt_i}{5A_e GS^2} + \frac{Vt_i}{A_e E_b}$$

E_b is the bulk modulus of the elastomer. The total vertical deflection of a laminated bearing is then the summation of the deflections of the individual layers. For a strip bearing, i.e. an unreinforced elastomeric bearing where the length is not less than 10 times the width, the effect of bulk modulus can be ignored so that

$$\delta = \frac{Vt}{5A_e GS^2}$$

These expressions may be used to estimate the change in deflection between one-third of the total load and full load, with an accuracy of the order of $\pm 25\%$.

The actual deflection of a bearing includes an initial bedding down phase that can produce deflections of approximately 2 mm. Thereafter, the stiffness of the bearing increases with increasing load. Where the vertical deflection under load is critical to the design of the structure, the stiffness of the bearing should be ascertained by tests. However, a variation of as much as $\pm 20\%$ from the observed mean value may still occur. When a number of similar bearings are used at a support and the differential stiffness between the bearings is critical for the structure, a variation of compressive stiffness should be allowed in the design equal to either $\pm 15\%$ of the value estimated or $\pm 15\%$ of the mean value observed in tests.

The calculations for the deflection of plain pad and strip bearings are likely to underestimate the deflection under permanent load and overestimate the deflection under transient loads.

In an elastomeric bearing, horizontal movement is accommodated by shear in the elastomer. The horizontal force resulting from this shear is given by the expression

$$H = \frac{AG\delta_r}{t_q}$$

where

A is the actual plan area of the elastomer slab;

G is the shear modulus of the elastomer;

δ_r is the maximum resultant horizontal displacement;

t_q is the total thickness of elastomer in shear.

Typical values of G are given in Table 4.6; allowances should be made for the variation of G due to temperature and dynamic loading.

Inclined bearings

A single bearing has identical shear characteristics in all horizontal directions, but an assembly of bearings can be stiffened in one direction by mounting them in pairs inclined towards each other. This is particularly useful where high lateral loading has to be resisted, because the resistance to normal longitudinal movements is unchanged. It is also useful in establishing an expansion centre line.

The effective lateral horizontal stiffness K_h of a pair of identical bearings of shear stiffness K_s and compressive stiffness K_c, if oppositely inclined each at an angle of θ degrees to the horizontal, is given by the expression:

$$K_h = 2(K_c \sin^2\theta + K_s \cos^2\theta)$$

Similarly, the effective vertical stiffness K_v is given by the expression:

$$K_v = 2(K_c \cos^2\theta + K_s \sin^2\theta)$$

To prevent lift off in one of the pair of bearings, it is necessary to ensure that

$$\frac{V}{H} > \frac{K_c \tan\theta}{K_c \tan^2\theta + K_s}$$

provided $\theta < 5°$, where V and H are the applied vertical and horizontal loads.

Where significant transverse rotation of the superstructure can occur at the bearings due to eccentric load effects, the effective value of K_h is significantly reduced to the following value:

$$K_h = \frac{2\,K_c K_s}{K_c \cos^2\theta + K_s \sin^2\theta}$$

However, excessive tensions may be developed in the bearings due to superstructure rotations, especially if the compressive deflections are not sufficient, and the overall performance of the bearings assembly should then be re-assessed from first principles.

Vibration isolation

For vibration isolation the natural frequency of the supported structure should be less than that of the disturbing frequency by a factor of at least 2 and preferably 3. The natural frequency η_f is given by

$$\eta_f = \frac{1}{2\pi}\left(\frac{K_s D}{V/g}\right)^{1/2}$$

where

K_s is the static stiffness of the bearing system;
D is the dynamic stiffening factor (Table 4.7);
V is the vertical load;
g is the acceleration due to gravity.

A method of designing elastomeric bearings with lead plugs for the seismic isolation of bridges is given in reference [23].

Creep in elastomers

Long-term loads cause time-dependent effects in elastomers through creep and stress relaxation. Creep effects are logarithmic functions of time and are dependent on temperature, stress history and elastomer formulation. Maximum compressive creep values of the order of 25% of the instantaneous deformation (including initial relaxations of the order of 10% followed by 5% per year) have been indicated by tests.

Generally, for bridge-bearing applications, the effects of creep and stress relaxation in the elastomer need not be considered since a significant proportion of the creep has taken place by the time deck joints are installed or final surface finishes applied: also shear relaxation effects generally reduce effects on piers and abutments.

4.5.7 POT BEARINGS (Figs 4.31–4.33)

The problem of providing a bridge bearing to take large vertical loads combined with a high degree of rotation cannot be overcome by using laminated rubber bearings; they would require large bearing surfaces to keep all sides of the bearing in compression and this would result in an uneven pressure distribution on the concrete, which in most cases would be too great. In the past, this problem was overcome by using heavy steel rocker bearings.

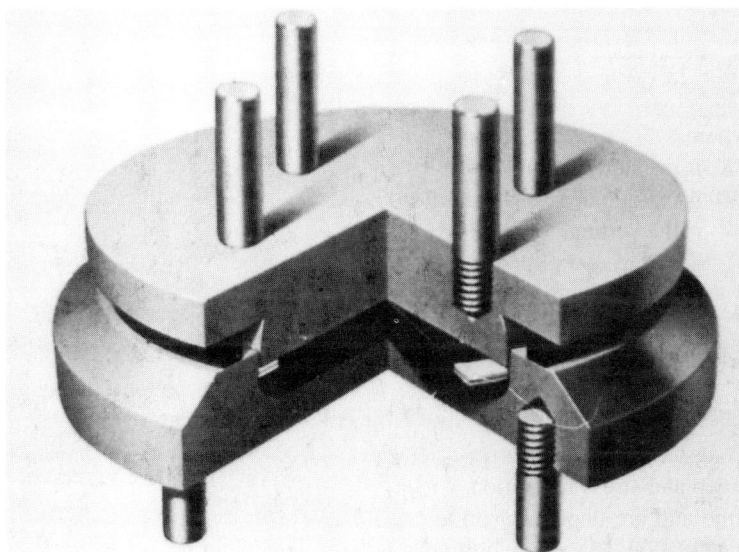

Fig. 4.31 The BTR Solarbridge floating piston bridge bearing incorporates a hydrostatically loaded captive rubber disc, allowing the upper members of the bearing to rotate about any axis, with negligible shift in centre of pressure. (Courtesy BTR Silvertown Ltd)

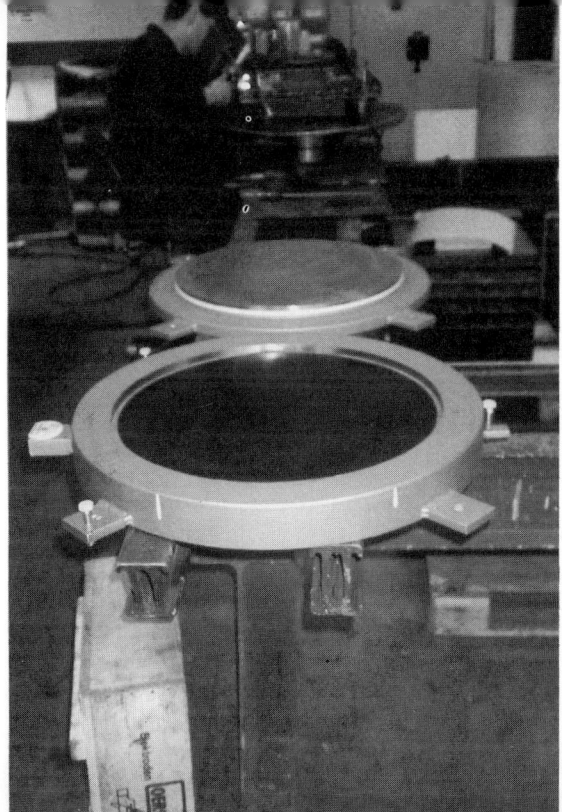

Fig. 4.32 Mageba pot bearings for the Störabelt project in Denmark. (Courtesy Kerby Engineering Co. Ltd.)

a

b

Fig. 4.33 Mageba pot bearings for the Neugut Viaduct, Switzerland. Vertical load 75 000 kN. (Courtesy Mageba S.A.), (a) before installation; (b) after installation.

The pot type of bearing, which was developed in 1959, consists of a circular non-reinforced natural rubber or neoprene pad, of relatively thin section, totally enclosed in a steel pot with the load applied to the elastomer via a piston attached to the upper bearing plate. A seal is used to prevent rubber extruding between piston and pot. The elastomer is thus prevented from bulging and has been shown by extensive testing to behave similarly to a fluid under high pressure. The result produces a bearing suitable for rotations of up to 1/50, at the same time giving the effect of a point-contact rocker bearing since the centre of pressure does not vary by more than 4%. This type of bearing can be combined with a sliding or roller bearing to give horizontal movement. Experience has shown that the pot bearing is a compact, economical and efficient bearing.

The stress in the elastomer in pot bearings due to the design load effects is limited by the effectiveness of the seal preventing the rubber from extruding between the piston and the pot wall, but it should not exceed 40 N/mm^2 at the serviceability limit state. The lateral pressure exerted on the confining cylinder walls resulting from vertical loading on the elastomeric pad can be considered to be that produced by the pad acting as a fluid.

The rotation of pot bearings about a horizontal axis should be limited so that the vertical strain induced at the perimeter of the elastomeric pad, at the serviceability limit state, does not exceed 0.15.

The thickness and hardness of the elastomer have a direct relationship with the resistance of pot bearings to rotation, as does the friction between the piston and pot. The latter is increased by increased force acting on the bearing. Sufficient test results should be available for a given elastomer stress, hardness and thickness to enable the resistance of the bearing to rotation to be calculated; otherwise prototype tests should be made.

The design of pot bearings is complex and best left to the specialist bearing manufacturers. At the allowable rubber stresses the steel stress in a normally designed pot is well below the yield limit and it is almost impossible to compute their magnitude with any accuracy. Buchler [9] has shown that the seating and any PTFE sliding element can have a considerable effect on the stresses in pot bearings. Tests [26, 27] have shown that for a properly designed pot bearing the supporting concrete structure is likely to fail before the bearing. Care must be taken to ensure uniform loading on pot bearings when used in steel structures.

4.5.8 FABREEKA BEARINGS

Fabreeka bearings (CCL Systems) employ a Fabreeka pad to take the vertical loads and allow for rotation (Fig. 4.34). Fabreeka consists of multiple layers of lightweight fabric, 0.275 kg per square metre, impregnated with a high quality natural rubber compound containing anti-oxidants and mildew inhibitors, and vulcanized into slabs of uniform thickness.

With a limiting side dimension of 600 mm, Fabreeka pads are manufactured in nominal thickness of 13 mm, 16 mm, 19 mm and 25 mm to any required length. Greater thicknesses are obtained by bonding standard thicknesses together. For pads thicker than 55 mm, a thin rigid metal substrate is introduced as an intermediate reinforcing layer. Fabreeka pads are manufactured to withstand compressive loads perpendicular to the plane of laminations of not less that 70 N/mm^2 before breakdown. In practice, a much lower value is used in the design of these bearings, thus providing a large factor of safety.

Tests have shown that Fabreeka pads withstand large horizontal shear stresses and shear strains up to 50% without failure. For design purposes it is recommended that a limiting value of 0.7 N/mm^2 be used for allowable shear stresses on a plain Fabreeka pad. When a Fabreeka pad is subjected to a sustained static load there is a gradual increase in the compression of the pad with the passage of time. The creep varies with the intensity of the stress and with the time the load has been sustained. The rate of creep decreases with the passage of time. Under a sustained static load of 20 N/mm^2 the creep on a Fabreeka pad was found to be under 3% of its original thickness, over a period of 192 h.

Horizontal forces on fixed end bearings are taken by dowels placed between the bearing pads although the Fabreeka pads can withstand a limited horizontal force themselves. Horizontal movement is catered for by a PTFE/stainless steel sliding surface. Unfilled PTFE is bonded to a rigid stainless steel substrate, which, in turn, is bonded to the upper surface of the Fabreeka pad.

Fig. 4.34 Fabreeka bearings. (Courtesy CCL Systems Ltd), (a) inspection before assembly; (b) guided bearing dismantled; (c) guided bearing assembled.

The sliding upper unit of the free end bearing comprises a corrosion protected mild steel backing plate faced with a thin stainless steel plate. The mild steel plate has a 2.5 mm overhang on all sides of the bearing with respect to the stainless steel plate.

Generally, the backing plate is made large enough to cater for all specified movements due to shrinkage, creep, temperature changes, etc. in both the transverse and longitudinal direction. In addition, the stainless steel plate is provided with a clear overhang of at least 10 mm beyond the PTFE on each side of the bearing at the extremes of movement.

The design of Fabreeka bearings is based on empirical data based on fatigue tests which indicate that under cyclic loading conditions the load-deflection characteristics of Fabreeka alter. The extent of the change is dependent upon the value of the load, number of loading cycles and the rate of the cyclic loading.

The tests showed that the compressive stiffness of the pad, in the stress range normally used in bearing design, increased rapidly after the first few cycles but reduced after the 15 625th cycle. The stress/strain relationship is shown in Fig. 4.35 and is used in the bearing design since it takes into account the highest edge stress which will occur due to rotation.

4.5.9 WABO–FYFE BEARING

The Wabo–Fyfe bearing was developed and patented internationally by Elastometal Limited of Canada, the first structural applications taking place in 1972. Rotation is accommodated by Bonafy structural elements incorporated in the bearing assembly as illustrated in Fig. 4.37.

Bonafy is a polyether urethene material developed in conjunction with E.I. Du Pont de Nemours & Co. (Inc.) specially for structural bearing applications. According to claims by the manufacturer, Bonafy is resistant to attack by oils, solvents, oxygen, ozone, moisture and sea water with a design operating range from –73°C to +121°C (–99°F to +250°F).

The specially shaped Bonafy element is retained in position by steel limiting rings fixed to the upper and lower bearing plates. Horizontal

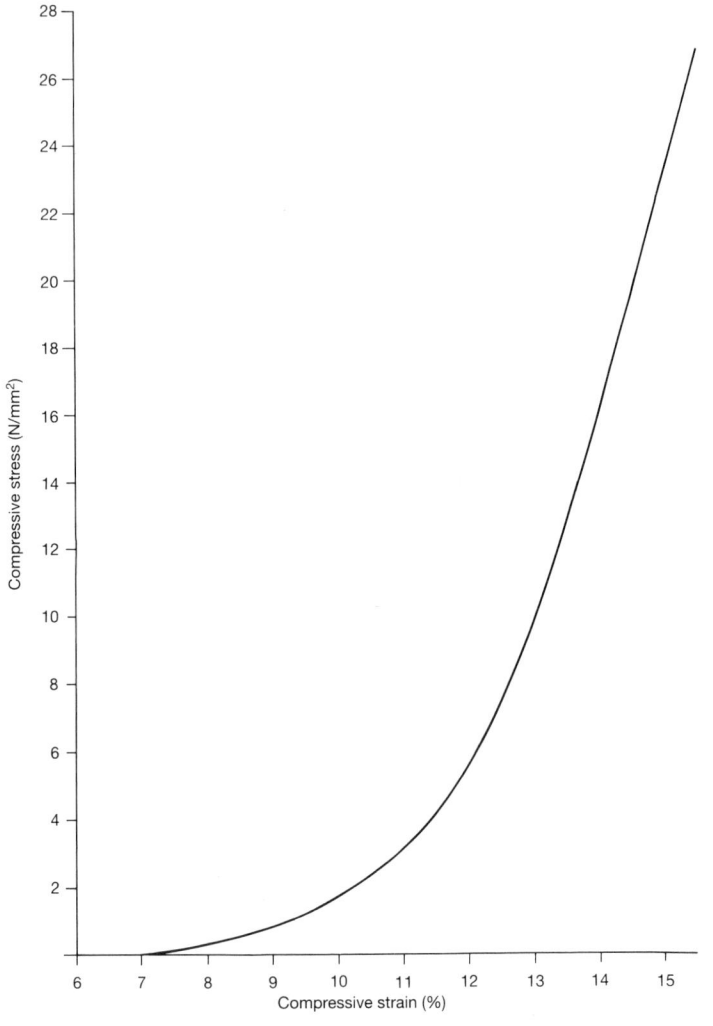

Fig. 4.35 100 000 cycle fatigue test on Fabreeka bearing. Stress/strain graph for 15 625th cycle. (Courtesy CCL Systems Ltd.)

Fig. 4.36 Wabo-Fyfe bearings. (Courtesy Watson Bowman Acme Inc.), (a) uplift bearing in the Calgary Saddledome, Canada; (b) placing a Wabo-Fyfe bearing on the Dufferin Expressway, Quebec City, Canada; (c) Wabo-Fyfe bearings in the Manhattan Bridge, New York, USA.

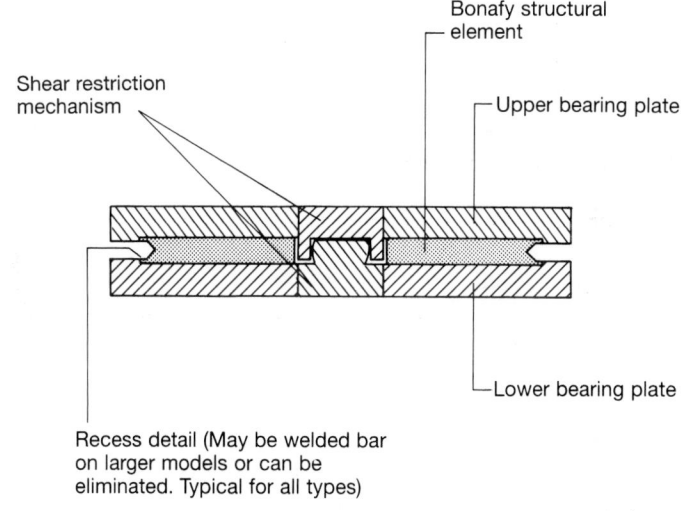

Fig. 4.37 Wabo-Fyfe Model II fixed bearing. (Courtesy Watson Bowman Acme Inc.)

shear forces are taken on a dowel pin mechanism in the centre of the bearing which transfers load directly between the upper and lower bearing plate thus eliminating shear on the Bonafy element and thereby allowing high loads to be taken by the material. Horizontal movement is catered for, as in the Fabreeka bearing, by stainless steel/ PTFE sliding unit on top of the bearing. This gives a compact bearing with high rotational capacity and ability to rotate at low loads.

The bearing is similar in some respects to a pot bearing, but the rotational element has no critical tolerances compared with pot bearings and no sealing rings or mating curved surfaces. The bearings can be designed to take uplift. The maximum working design compressive stress on the Bonafy is 35 N/mm^2 and the long-term deflection of the Bonafy structural element is stated as 7% of the element thickness.

Fig. 4.39 BTR Solarbridge uplift bearing capable of horizontal low-friction movements together with rotation, under compression or uplift loads. (Courtesy BTR Silvertown Ltd.)

Fig. 4.38 Mageba uplift bearing. (Courtesy Mageba S.A.)

Fig. 4.40 BTR Solarbridge universal pin bearing which resists horizontal forces only and permits rotation about all axes. (Courtesy BTR Silvertown Ltd.)

Fig. 4.41 Manufacture of Mageba type TE pot bearings. (Courtesy Mageba S.A.), (a) machining the bearing; (b) piston lead for pot bearing showing guide bar; (c) machining a 1.8 m diameter piston.

4.5.10   GUIDES

It is often necessary to constrain the movements of bearings in a particular direction, e.g. the central bearings of a bridge may be constrained to move only along the line of the bridge to prevent the whole bridge working off the bearings transversely. In these cases movement in the desired direction is ensured by the use of guides. They may take the form of an independent guide bearing or form part of a bearing performing other functions.

Two broad classes of guides can be recognized:

1. For normal use, coming into contact as a result of external loading, for example, due to wind or traffic centrifugal forces.

Compactness is the important criterion and somewhat higher friction along the guide is acceptable, because it adds very little to the total resistance to movement in the bearing.

To reduce friction the guides may be faced with PTFE. In order to allow higher bearing stresses to be employed, and to improve wear and deformation characteristics, fillers may be incorporated in the PTFE. However, in some guides which are constantly in contact, for example in circular structures, a lower coefficient of friction may be required. In this case, the sliding surfaces should be based on virgin PTFE and stainless steel, designed in accordance with the rules given for the principal sliding surfaces.

Commonly used materials for facing guides are:
1. unfilled PTFE;
2. PTFE filled with up to 25% by mass of glass fibres;
3. lead-filled PTFE in a bronze matrix;
4. PTFE reinforced with a metal mesh.

It is essential that all PTFE should be securely attached to the guides; reliance should not be placed on bonding alone for pure PTFE.

2. For emergency use, provided solely to limit movement, for example, due to accident or subsidence.

In this case the resistance to movement along the guide is immaterial, so that metal-to-metal contact is acceptable. The metal should be corrosion resistant.

The frictional resistance to movement of guides should either be significantly smaller than that of the main bearing or the resulting frictional effects taken into account in the bridge design.

Under serviceability design load effects, the average pressure on glass-filled PTFE in guides should not exceed 45 N/mm^2, and on PTFE in a metal matrix 60 N/mm^2. Permissible values for other PTFE materials should be established by tests. If bronze is used for guides, the contact bearing stress should not exceed 30 N/mm^2.

4.5.11   CONCRETE HINGES

The Freyssinet hinge

The Freyssinet hinge (Fig. 4.42) allows rotation of a member to occur by providing a thin flexible throat within the member where large rotations can occur while only small moments are transmitted to the structure. The size of the throat is normally only a fraction of the size of the member, but it is able to carry very high axial and shear forces because of the biaxial or triaxial restraining influence of the surrounding concrete. It has been found experimentally that compressive stresses many times the value of the cube stress of the concrete can be applied to a restrained concrete throat without causing crushing, and indeed the primary factor that has to be considered in the design of these hinges is not the compressive stress in the throat but the bursting tension either side of the throat.

In the design of Freyssinet hinges the following basic assumptions are made.
1. The effect of any reinforcing steel which may be incorporated in the throat of a hinge for ease of handling is neglected.
2. The effect of shrinkage cracks in the throat is neglected.
3. For short-term loading the behaviour of the concrete is elastic.
4. For long-term loading the creep is proportional to the initial stress.
5. In considering the transverse tensile forces in the hinge the tensile strength of the concrete is neglected.
6. Rotation takes place in one plane only.

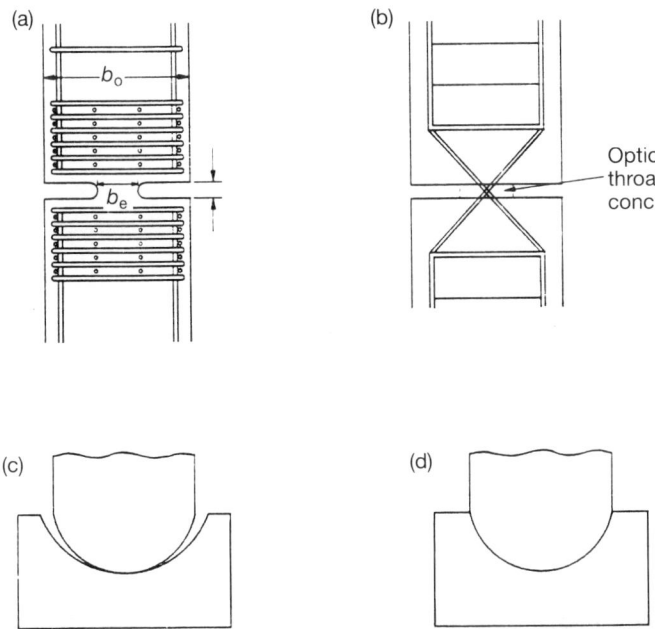

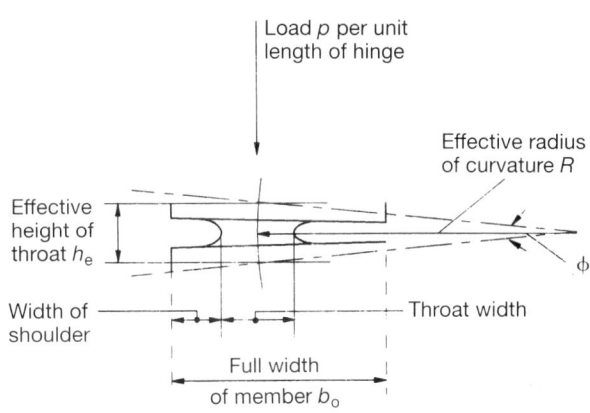

Fig. 4.43   Freyssinet hinge.

Fig. 4.42   Types of concrete hinge (a) Freyssinet hinge; (b) Mesnager hinge; (c) saddle bearing; (d) sliding hinge.

Stress distribution in throat

Stresses due to axial load

When the throat of a concrete hinge (Fig. 4.43) is subjected to an axial load, stress concentrations occur at the edge of the throat because of the convergence of the stress trajectories in this region. The stress distribution across the throat of an axial compressive load is the form shown in Fig. 4.44, the maximum stresses being at the sides of the throat. Tests have shown that if this maximum stress is defined as N_p/b_e then the notch factor N may be taken to be 1.5 for the range of tests carried out (i.e. for throat widths up to 200 mm). In no tests carried out

so far has a compressive failure of the throat been achieved, but it is recommended that the average compressive stress in a throat should be limited to twice the cube strength or 105 N/mm^2 (15000 lbf/in^2) whichever is lower.

Stresses due to rotation

In spite of the high compressive stresses that occur within the throat of a hinge, the triaxial restraint of the surrounding concrete maintains the throat in an elastic state. The bending stresses across the throat are therefore linear and may be superimposed upon the non-linear axial stresses as shown in Fig. 4.44.

Experimental work has shown that the behaviour of the concrete is elastic for short-term loading and that under long-term loading the creep is proportional to the initial stress. It is therefore possible to calculate from elastic considerations the rotation ϕ which, rapidly applied, would just cause annulment of the compression on one side of the throat. Tensile stresses in the throat must not be permitted except for shrinkage stresses which may arise during construction. As the throat of the hinge can be considered to remain elastic during rotation,

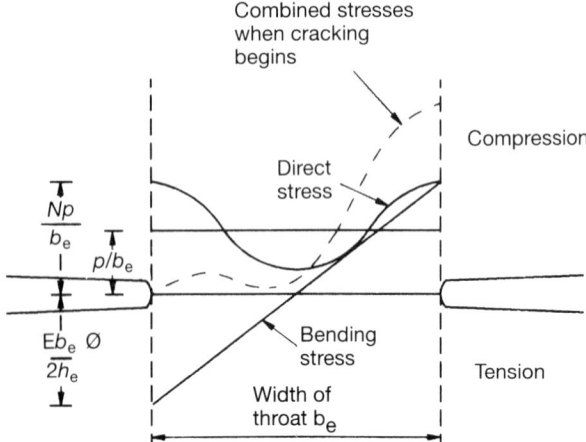

Combined stresses
when cracking
begins

Compression

Direct
stress

$\dfrac{Np}{b_e}$

p/b_e

$\dfrac{Eb_e\,\varnothing}{2h_e}$

Bending
stress

Tension

Width of
throat b_e

Fig. 4.44   Stresses in throat of Freyssinet hinge.

it can be shown that the moment required to rotate the hinge through an angle ϕ is

$$M = \frac{E\phi b_e^3}{12h_e}$$

where h_e is the effective height of the hinge.

The extreme fibre bending stress due to a superimposed moment on the hinge

$$f_b = \pm \frac{Eb_e\phi}{2h_e}$$

Therefore, for no tensile stresses in the throat

$$\frac{Np}{b_e} \geqslant \frac{b_e}{2h_e}\sum E\phi$$

where $\Sigma E\phi$ represents the sum, at any time, of the products of rotation and modulus of elasticity for each type of loading. It has been found experimentally that h_e, the effective height of the throat, can be taken to be 125 mm, and hence the above expression may be re-written as

$$b_e \geqslant \left(\frac{375p}{\sum E\phi}\right)^{1/2}$$

The value of the modulus of elasticity of the hinge for long-term rotations due to shrinkage, creep, elastic shortening and permanent loads may be taken as being half that for temperature and transient loads.

If E is the modulus of elasticity for temperature and transient loads, ϕ_S is the total rotation due to transient loads and temperature and ϕ_L is the total rotation due to shrinkage creep, elastic shortening and permanent loads, then

$$E\phi = E\left(\phi_S + \frac{\phi_L}{2}\right) = E\phi_e$$

Values of the modulus of elasticity of concrete under short-term loading are given in Table 2 of BS 5600: Part 4 [28].

The maximum value of rotation per unit axial load often occurs under the most lightly loaded condition and should be determined for each condition of loading.

Notches

Although it has been shown that very high compressive stresses can be carried by the throat of the hinge when triaxially restrained, spalling will occur at the ends of the hinge unless the throat is adequately recessed there. If the notches forming the throat are rectangular in cross-section, spalling will also occur at the corners. Ideally the shape of the notch in the immediate vicinity of the throat should approximate a parabola as shown in Fig. 4.45. The curve should merge into parallel straight lines which continue to the edges of the member. (A divergence of up to 1 in 20 may be allowed to facilitate easy withdrawal of the formwork.) The width of the throat b_e should not be less than 50 mm and each end of the throat should be recessed at least 75 mm from the face of the member to prevent spalling of concrete under load.

Values for maximum unit compressive load p and equivalent rotation ϕ_e are given in Tables 4.8 and 4.9 for two typical characteristic concrete strengths.

Bursting tensions

The tensile stresses due to bursting are paramount in the design of Freyssinet hinges. They occur in members adjacent to both sides of the hinge over a length equal to the width of the member. They may be calculated by conventional end-block analysis. Extensive tests carried out by the Cement and Concrete Association have shown that the total area of steel required to resist bursting is

$$\frac{R}{f_y} \quad \text{in the transverse direction}$$

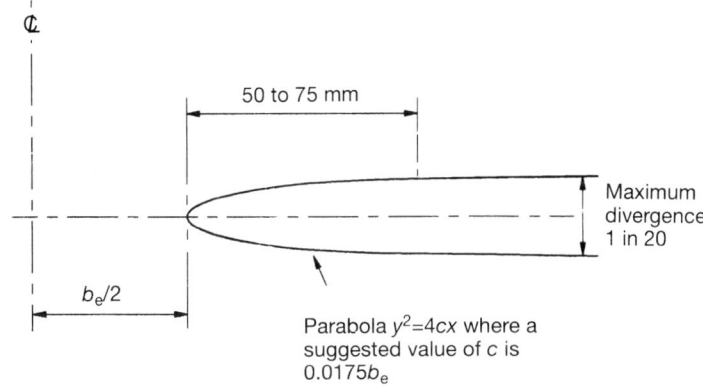

Fig. 4.45 Typical shape of Freyssinet hinge notch.

Table 4.8 Maximum load and rotations for various throat widths with concrete having a characteristic strength of 45 N/mm² (E = 32.5 kN/mm²)

Throat width b_e (mm)	Maximum axial compressive load p_{max}/unit length of throat (N/mm)	Maximum permissible value of $\frac{\phi_e}{p}$ (rad/(N/mm)) × 10⁻⁸
50	4200	475
62.5	5250	305
75	6300	210
87.5	7350	155
100	8400	120
112.5	9450	90
125	10500	75

Equivalent rotation ϕ_e = (rotation due to live loads and temperature)

+ $\frac{1}{2}$ (rotation due to shrinkage, creep, elastic shortening and dead load).

Table 4.9 Maximum loads and rotations for various throat widths with concrete having a characteristic strength of 52.5 N/mm² (E = 34.5 kN/mm²)

Throat width b_e (mm)	Maximum axial compressive load p_{max}/unit length of throat (N/mm)	Maximum permissible value of $\frac{\phi_e}{p}$ (rad/(N/mm)) × 10⁻⁸
50	5250	440
62.5	6650	280
75	7900	195
87.5	9200	145
100	10500	110

Equivalent rotation ϕ_e = (rotation due to live loads and temperature)

+ $\frac{1}{2}$ (rotation due to shrinkage, creep, elastic shortening and dead load).

$\frac{R}{3f_y}$ in the longitudinal direction

where R is total resolved component of force acting on hinge:

$$R = L\left[p^2 + q^2\right]^{\frac{1}{2}} \text{ (N)}$$

where p is the axial load per millimetre applied to hinge (N/mm), q is the shear force per millimetre applied to hinge (N/mm) and L is the length of hinge perpendicular to direction of rotation (mm).

This reinforcement should all be placed within the zone of bursting and as near to the throat as possible. Bond is normally achieved by welding the reinforcement in a series of mats. The mats should be detailed so as to permit the proper compaction of the concrete and the distance between bars sensibly parallel should nowhere be less than 20 mm. Steel stresses in the transverse mat reinforcement should not exceed 105 N/mm² at the serviceability limit.

Shear across throat

The large compressive stresses in the throat of Freyssinet hinges give them a high shear resistance. The principal tensions in the throat can be calculated by conventional methods.

The practice of placing shear steel in the throat has been discovered to be of little value. In many cases, in fact, it can have an adverse effect on the hinge.

Where it is required to safeguard the structure against some shear loading of an unexpectedly high intensity (vehicle impact, for instance), dowelling of the member in some position remote from the hinge is preferable to passing shear dowels through the throat.

Prestressed hinges

Uplift may never be permitted in concrete hinges. Where there is danger of uplift, it is possible to prestress across the hinge. The prestressing steel should again be kept away from the throat when possible.

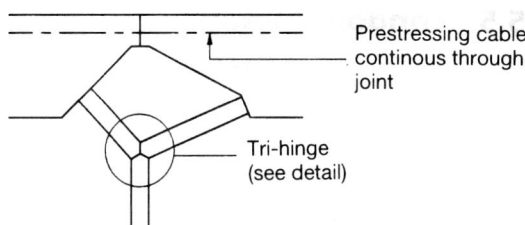

a

b

Fig. 4.46 Multiple concrete hinges (a) rhomboid joint; (b) detail of tri-hinge.

Mesnager hinge

The Mesnager hinge (Fig. 4.42(b)) is a forbear of the Freyssinet hinge. It carries all the load on reinforcement passing through its throat. In many cases the throat concrete is omitted altogether. The capacity of the Mesnager hinge is considerably less than that of the Freyssinet hinge which has now virtually superseded it.

Saddle bearings

Saddle bearings (Fig. 4.42(c)) are used when it is required to accommodate high rotations and shear forces in conjunction with relatively low axial forces. The bearings have the same high resistance to crushing as Freyssinet bearings, but they experience the same bursting tendency and should be reinforced with the same amount of steel. For design purposes the ratio of contact width b_e to total width b_0 may be taken as 0.2. The resistance to rotation of a saddle bearing is small, although the effect of the displacement of the point of contact when the bearing rotates must be considered.

Sliding hinges

Sliding hinges (Fig. 4.42(d)) have a similar function to saddle bearings. They are easier to form than saddle bearings but may transmit a significant moment. If the friction coefficient between the sliding faces is μ and the radius is r, then the moment transmitted by the hinge is given by $M = Rr\mu$ (where R is the resultant force acting on the hinge). No rotation of the hinge will occur until this moment has been developed.

Tri-hinge

The tri-hinge shown in Fig. 4.46(b) is a special application of the Freyssinet hinge. It has a better performance than the conventional Freyssinet hinge because of the multidirectional loading imposed on it.

The tri-hinge is used in conjunction with the rhomboid joint shown in Fig. 4.46(a). This joint (which is a concrete version of the Wichert truss support) allows a substantial settlement to occur at a column of an effectively continuous structure, without any change in continuity moment, because the system is statically determinate.

4.6 Selection of bearing type

4.6.1 INTRODUCTION

The designer has to distinguish clearly between the characteristics of bearings of various types and evaluate correctly their parameters as part of the structure.

Elastomeric bearings have a force–displacement relationship which generally enables them to be included in the elastic analysis of the structure. For roller bearings, the force required to set the roller in motion has to be considered first; once it is rolling a nominally constant friction force is involved.

For sliding bearings, a somewhat similar set of circumstances applies. Movement of the structure, if small and below the frictional resistance of the bearing, is absorbed by the elasticity of the structure. Thereafter, a greater movement builds up a force equal to the sliding friction and the force then remains substantially constant, although dependent on such factors as speed of sliding and contact stress.

For modern sophisticated structures, it may not be valid to change the type of bearing or even the manufacture without considering the effect on the rest of the structure.

Modern bridge designs require the simultaneous treatment of both transverse and longitudinal movements. For bridges of varying width, skew and curvature, care must be taken to ensure that the movements to be expected are properly evaluated to prevent overstressing the bearing. In certain circumstances when large movements are involved, it may be preferable to use sliding bearings which can allow unrestrained movement in certain directions and inhibit movement in other directions by guide plates.

The types of bearings for a particular structure will depend upon the functions they are called upon to perform. These in turn will depend upon whether the deck is simply supported or continuous, the flexibility of the substructure and the ambient climatic conditions. The bearings may be required to constrain the structure in a prescribed manner. This is particularly true for bridges curved in plan. If a bridge contains a number of bearings at a point of lateral restraint and particu-

larly if they are spaced far apart, only one bearing in the line should be fixed with respect to transverse movement whilst others should permit movement in all directions. For inclined superstructures, fixed bearings should preferably be located at the lower end of the structure. It is also necessary to ensure that there is compatibility between bearing and deck joint movements in the vertical and transverse directions. The suitability of various types of bearing for differing functions are summarized in Table 4.10 and detailed considerations are given below.

For example, British Rail tend to use rubber strip under precast prestressed concrete beams since railway loading precludes the use of long prestressed beams. For short-span steel girders one end is fixed and the other is left free without any special bearing, since vibration allows expansion movement to take place. For large spans, use is made of special bearings as supplied by Glacier, which have been found very reliable.

4.6.2 HORIZONTAL MOVEMENTS

The limit of horizontal movement from all causes in elastomeric bearings is related to their depth. This is not so with mechanical bearings which can have almost unlimited capacity. The resistance of bearings to horizontal movement is important, particularly at the top of slender piers, where it can influence the entire pier design. The resistance of most mechanical bearings is frictional in character, being dependent upon, and approximately in proportion to the vertical loading. It is usually affected only to a minor extent by the speed, and not at all by the magnitude of the horizontal movement. The resistance of elastomeric bearings increases approximately linearly with horizontal movement, and is affected only to a minor extent by the magnitude of the vertical loading and the speed of the movement. With horizontal movements, up to about 20 mm, it is probable that the smallest resistance to movement would be obtained from elastomeric bearings. The compressibility and hence vertical deflection of elastomeric bearings under full load is greater than that of mechanical bearings, and this deflection increases with time, due to creep. However, once installed, only the live load deflection need be considered.

4.6.3 ROTATIONS

With any appreciable loading, it is difficult to design stable elastomeric bearings which accommodate more than approximately 0.02 radians rotation about a horizontal axis, whereas some mechanical bearings can accept 0.05 radians or more. The centre of rotation of a bearing employing an elastomer normally lies on the uppermost surface of the elastomer and the moment of resistance increases with the angle of rotation. In certain types of mechanical bearings, because of their geometry, the centre of the rotation is some distance away from the bearing and the moment of resistance is roughly constant, regardless of the angle of rotation.

When a structure rolls on a curved surface of radius R through an angle θ the point of load application moves a distance equal to $R\theta$ relative to the corresponding point on the substructure (Fig. 4.47).

Table 4.10 Bearing facilities

Type of bearing	Maximum loading kN[1]			Maximum translation (mm)[1]		Maximum rotation (rad)[1]			Seismic performance	Maintenance requirements	Typical applications[2]			
	Vert.	Long.	Trans.	Long.	Trans.	Long.	Trans.	Plan			Straight	Curved	Steel	Concrete
Roller	16 000	Nil	400	Unlimited	Nil	±0.05	Nil	Nil	Poor	Some	✔	–	✔	–
Rocker														
Linear	15 000	1500	1500	Nil	Nil	±0.05	Nil	Nil	Poor	Some	✔	–	✔	–
Point	20 000	1000	1000	Nil	Nil	±0.01	±0.01	Unlimited	Poor	Some	✔	–	✔	–
Knuckle														
Pin	25/mm	2.5/mm	–[3]	Nil	Nil	±0.05	Nil	Nil	Poor	Some	✔	–	✔	–
Leaf	12/mm	12/mm		Nil	Nil	±0.09	Nil	Nil	Fair	Some	✔	–	✔	–
Hinge[4]														
Saddle	6/mm	3/mm	–[3]	Nil	Nil	±0.09	Nil	Nil	Poor	Some	✔	–	–	✔
Freyssinet	10/mm	2.5/mm	2.5/mm	Nil	Nil	–[5]	Nil	Nil	Fair	None	✔	–	–	✔
Sliding														
Plane	3000	–[3]	–[3]	Unlimited	Unlimited	Nil	Nil	Unlimited	Good	Minimal	✔	✔	✔	✔
Cylindrical	15 000	1500	–[3]	Nil	Unlimited	±0.03	Nil	Nil	Good	Minimal	✔	–	✔	✔
Spherical	30 000	3000	3000	Nil	Nil	±0.05	±0.05	Unlimited	Good	Minimal	✔	✔	✔	✔
Pot	50 000	2500	2500	Nil	Nil	±0.01	±0.01	Nil	Good	Minimal	✔	✔	✔	✔
Disc	45 000	4500	4500	Nil	Nil	±0.04	±0.04	Nil	Good	Minimal	✔	✔	✔	✔
Elastomeric														
Unreinforced	1500	–[3]	–[3]	12	12	–[5]	–[5]	Small	Good	None	✔	✔	–	✔
Laminated	5000	–[3]	–[3]	50	50	–[5]	–[5]	Small	Good	None	✔	✔	✔	✔
Fabric	1000	–[3]	–[3]	Nil	Nil	±0.01	±0.01	Nil	Good	Minimal	✔	–	–	✔

[1] The maximum values quoted are those for bearings normally available to manufacturers' standard designs. It may not be possible for a bearing to achieve maximum capacity in all modes simultaneously.
[2] –: not suitable.
[3] Special arrangements required to prevent lateral movement and to take horizontal loads.
[4] Mesnager hinge not included as generally superseded by Freyssinet hinge.
[5] Maximum rotation depends on vertical load and dimensions of bearing.

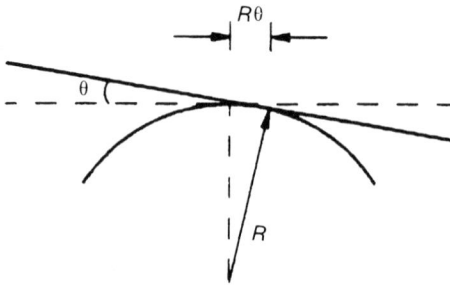

Fig. 4.47 Effect of rotation on point of support.

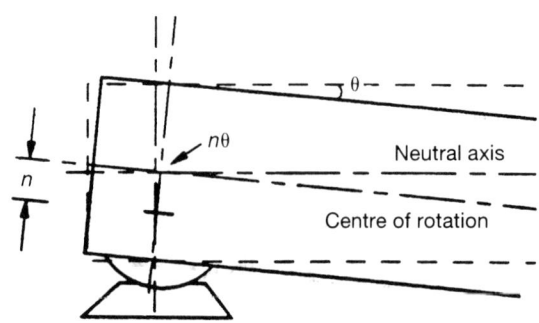

a

Unless the centre of the rotation of a mechanical bearing coincides with the neutral axis of the member it supports, the rotation of a free bearing through an angle θ will cause a relative movement of $n\theta$ on the sliding surface, where n is the distance between the neutral axis and the centre of rotation (Fig. 4.48). In a fixed bearing this relative movement cannot occur and so the member is moved bodily a distance $n\theta$ horizontally as a result of the rotation.

If the centre of pressure is to remain constant on the substructure (which may be important in the case of piers or columns), the moving element should be attached to the superstructure. In the case of curved sliding surfaces the shift in the centre of pressure due to frictional resistance is equal to μR, where μ is the coefficient of friction, but if R is small and with the low friction characteristics of modern sliding elements this shift will be small. The moment transmitted is $P\mu R$ where P is the resultant force acting on the bearing or hinge. No rotation takes place until this moment has been developed. Frictional resistance gives rise to horizontal forces, but these are independent of the magnitude of the sliding movement.

The magnitude of the shift of the centre of pressure on an elastomeric pot bearing depends upon a number of factors, including the characteristics of the elastomer and bearing geometry but is roughly proportional to the angle of rotation and does not normally exceed 5% of the radius of the elastomeric disc. Load eccentricity due to rotation

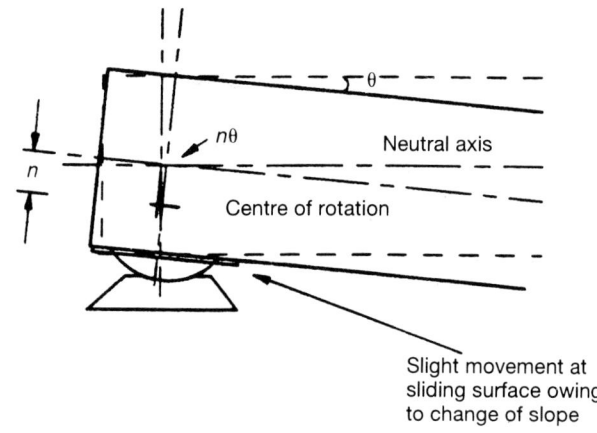

b

Fig. 4.48 Effect of rotation at (a) fixed and (b) free bearings.

of an elastomeric disc bearing is also roughly proportional to the rotation angle, but is greater than that of a pot bearing due to the need for the unconfined elastomer to be harder to achieve an acceptable load carrying capacity. Vertical load eccentricities for various types of bearing under rotation are summarized in Figs 4.49 and 4.50.

Correctly selected mechanical bearings can accept much greater rotations in plan, than elastomeric bearings.

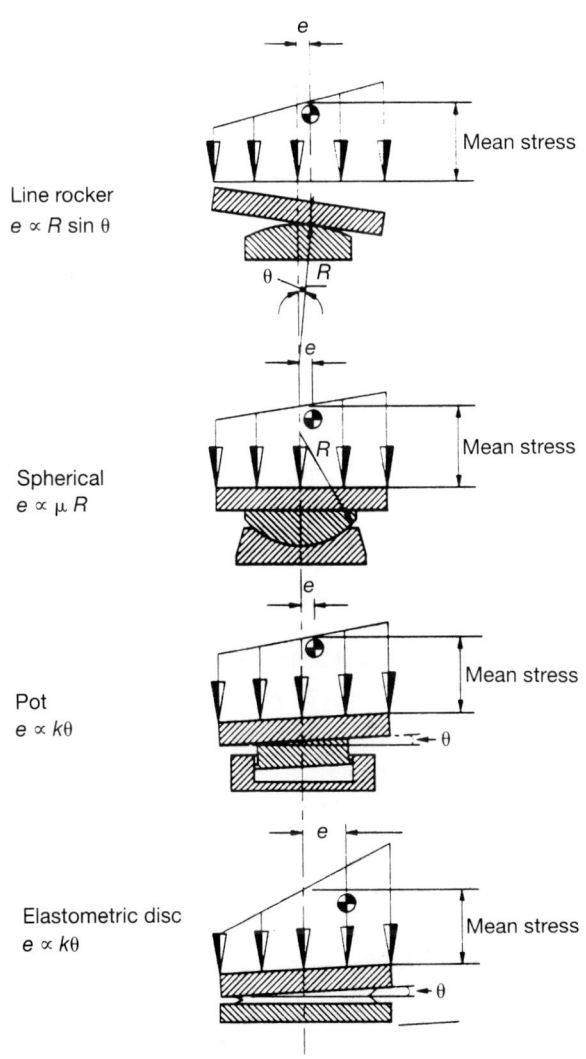

Fig. 4.49 Relative eccentricities for various types of bearings. (Courtesy The Glacier Metal Co. Ltd.)

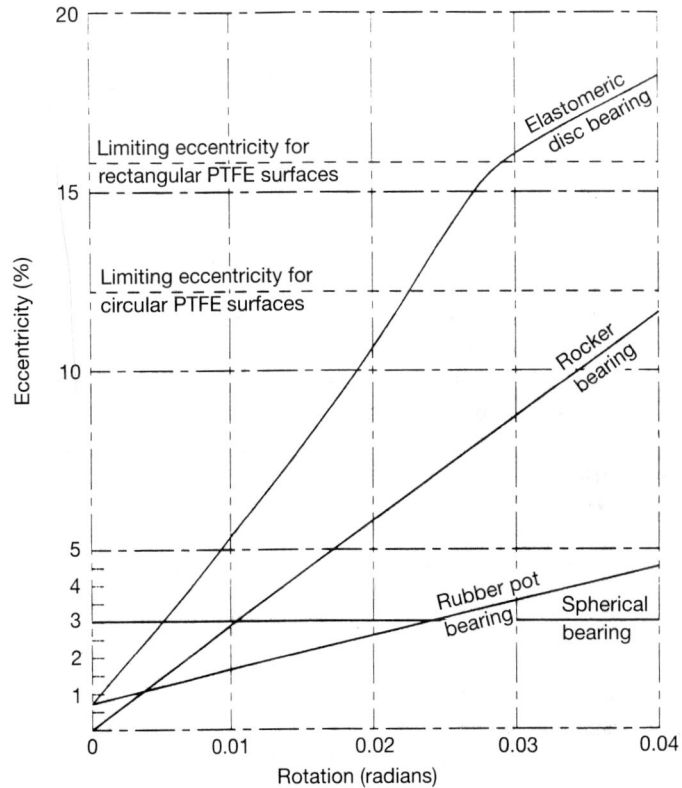

Fig. 4.50 Comparative eccentricities of bearing types. (Courtesy The Glacier Metal Co. Ltd.)

4.6.4   PLAN AREA

In theory at least, bearings can be made to any size, but bearings from manufacturers' standard ranges are likely to be cheaper than those specially designed.

Contact stresses under bearings should be related to the strength of the adjacent structure. The contact stress under elastomeric bearings are generally low, and seldom require special consideration, except possibly when bearings are supported on concrete plinths of approximately the same plan size as the bearings. With mechanical bearings the contact stresses can be much greater, and details in the vicinity of the bearings will almost certainly require special attention, for example, bearing stiffeners will be needed in steel members, and punching shear reinforcement or anti-bursting steel in concrete members.

4.6.5   LIFE AND MAINTENANCE

Properly designed bearings should last the life of the bridge. Elastomeric bearings need no maintenance, but may have a limited useful life, depending upon their environment. Most elastomers deteriorate steadily with time, and are subject to attack by materials such as oil, oxygen and ozone, and most are unsuitable for use at extremes of temperature.

Mechanical bearings require inspection and maintenance, so that access must be provided where appropriate, for example, for painting and greasing. Care should be exercised in using steel members against mechanical bearings of a dissimilar metal to ensure that cathodic action cannot occur, since this can lead to very rapid corrosion.

4.6.6   ROLLER BEARINGS

Roller bearings can be used in situations where no appreciable tilting movement takes place transverse to the centre line of the bridge. This is the case, for example, with straight bridges having bearings closely spaced in the transverse direction, the bridge structure incorporating rigid cross girders or diaphragms over the bearings and the supporting structure possessing substantial transverse rigidity. The direction of rolling of roller bearings should be perpendicular to the axis of rotation. Hence these bearings are unsuitable for skew bridges.

Modern roller bearings have very low friction characteristics, but movement causes the centre of pressure to move a distance equal to half that of the movement of the structure (Fig. 4.51(a)). This can have particular relevance to such features as web stiffening.

4.6.7   ROCKER BEARINGS

Linear rocker bearings are complementary to roller bearings and are used in similar circumstances at the fixed end of a bridge. Point rocker bearings are suitable in situations where rotation takes place about more than one axis or where the rotational axis is not perpendicular to the direction of translation. Horizontal movement can be accommodated by providing sliding elements in conjunction with rocker bearings. Load eccentricity is proportional to the rotation angle and radius of the contact surface.

4.6.8   KNUCKLE BEARINGS

Knuckle pin bearings are used where high longitudinal forces are involved, but like roller bearings they will not accommodate rotation about the bridge centre line. Leaf bearings are suitable where uplift is anticipated. Knuckle bearings are used in a manner similar to linear rocker bearings.

Roller and knuckle bearings have relatively high construction depths in comparison with other types of bearing so that inspection and maintenance are easier.

4.6.9   SLIDING BEARINGS

A big advantage that PTFE sliding bearings have over bearings of the roller type is that eccentricity of loading on the substructure can be obviated. Older types of sliding bearings with metal-to-metal contact may require individual assessment of their eccentricity. It can be seen from Fig. 4.51 that movement of a roller bearing inevitably gives rise

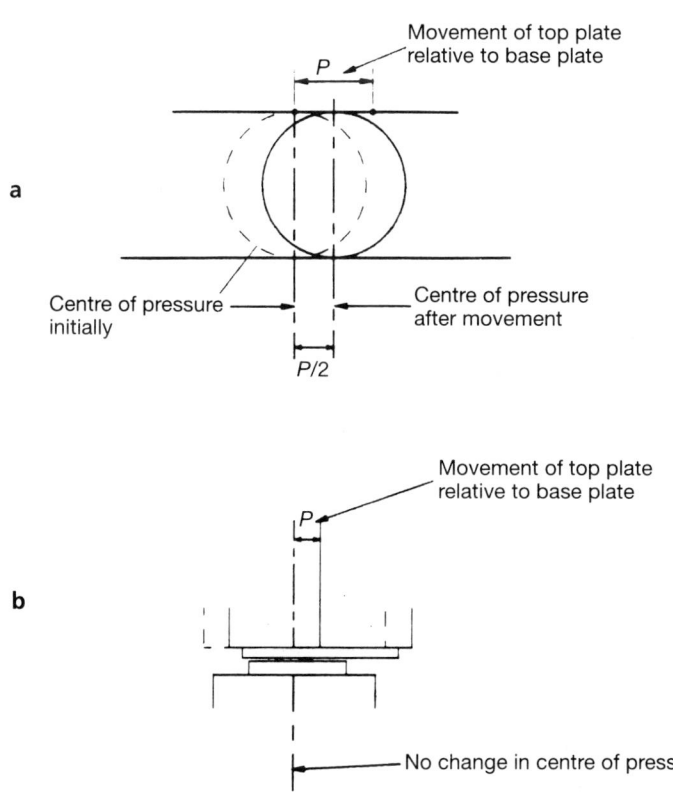

a

b

c

Fig. 4.51 Effect of translation on (a) roller bearing; (b) sliding bearing; (c) elastomeric bearing.

to an eccentricity of load from the mean position of half the movement. On the other hand, if the sliding plate is built into the moving structure and the rotational element is built into the substructure, the loading remains axial in spite of movement.

Plane sliding bearings can accommodate translational movement in more than one direction in addition to rotation in plan. Thus sliding bearings are particularly suitable for bridges curved in plan. Load eccentricities on PTFE bearing surfaces should not exceed one-eighth of the diameter for circular areas nor one-sixth of the side for rectangular areas to ensure that surfaces are not separated and the PTFE squeezed out of the unstressed periphery nor dirt admitted to the interface.

Cylindrical sliding bearings are used in a manner similar to linear rocker bearings, but will also permit translation parallel to the rotational axis. Spherical sliding bearings can be used in circumstances similar to those applicable to point rocker bearings. Although friction causes a shift in the centre of pressure, curved sliding bearings have the advantage that the eccentricity returns to zero once rotation has taken place.

Sliding bearings share with pot bearings the features of low construction height. Cylindrical and spherical bearings can also cope with large tilting movements.

4.6.10 POT BEARINGS

Pot bearings can be used in circumstances similar to those pertaining to point rocker bearings.

Load eccentricity is generally less with pot bearings than with other types. However, at low temperatures the rubber stiffens so that resistance to rotation increases. As with rocker and spherical bearings, horizontal movement can be accommodated by the addition of a plane sliding element.

4.6.11 DISC BEARING

Disc bearings are similar in action to pot bearings, except that the elastomer is unconfined. They are more economical than steel spherical bearings unless a high load capacity is required.

4.6.12   ELASTOMERIC BEARINGS

Elastomeric bearings cannot accept such large movements nor carry such heavy loads as other types of bearing. For larger movements they can be used in conjunction with sliding elements. Elastomeric bearings accommodate horizontal movement by shearing with a resultant horizontal reaction which is proportional to the movement (Fig. 4.51(c)). Elastomeric bearings must not be used in tension and, as they rely on flexing to take rotation, the maximum permissible rotation is dependent upon the vertical load for any given size of bearing to avoid this. Since they are very flexible horizontally, elastomeric bearings cannot accept large horizontal forces without some external aids such as dowels. They are, however, maintenance free and easy to install. In situations where elastomeric or mechanical bearings are equally suitable, the elastomeric type will generally prove to be the cheaper. If used on a steel bridge care must be taken to ensure that the combined thickness of the flange and sole plates is sufficient to ensure uniform loading on the elastomeric bearing. For slab type bridges with small horizontal movements, plain rubber bearing strips are suitable. Plain rubber pads should be restricted to fixed ends or expansion ends of simple structures not exceeding 10 m in span or width. Elastomeric bearings are useful for damping vibrations.

4.6.13   CONCRETE HINGES

Concrete hinges essentially carry direct load and, with the possible exception of Mesnager type hinges, have limited capacity for transverse loads and will not resist uplift forces. Thus they are mainly used at the feet of arches or portal frames. Increased horizontal loads can be taken by dowels or by prestressing the hinge, which method can also be used to overcome uplift. A short column with a hinge at each end will accommodate horizontal movements. Saddles can accommodate large rotations, but the shift of the centre of pressure and hence the resulting eccentric bending moment is correspondingly large (section 4.6.3). High quality concrete and workmanship are essential for concrete hinges.

4.6.14   THRUST BEARINGS

Special bearings may be required to transmit large horizontal forces whilst only carrying light vertical loads (e.g. bridges curved in plan or subject to earthquake forces) or to take uplift. These can take various forms, but horizontal movement is usually catered for by sliding members. Temporary stops can be fitted to normal sliding bearings, if necessary, to deal with lateral forces arising during a limited period (e.g. during construction). Other applications include bearings for arch ribs or pressurized pipes.

4.6.15   SYMBOLIC REPRESENTATION

Symbols to indicate the movement potential of various types of bearings are given in Table 4.11.

Table 4.11   Symbolic representation of bearing facilities

Symbol	Function	Suitable type of bearing
◯	All translation fixed. Rotation all round	Point rocker bearing Pot bearing Fixed elastomeric bearing Spherical bearing Compound cylindrical bearing
◄─◯─►	Horizontal movement constrained in one direction only. Rotation all round	Constrained point rocker sliding bearing Constrained pot sliding bearing Constrained elastomeric bearing Constrained spherical sliding bearing Constrained compound cylindrical bearing
◄─◯─► (with vertical arrows)	Horizontal movement in all directions. Rotation all round	Free point rocker sliding bearing Free pot sliding bearing Free elastomeric bearing Free spherical sliding bearing Free compound cylindrical bearing
◄═══►	Movement constrained in one direction only. No vertical load	Guide bearing
▭	All translation fixed. Rotation about one axis only	Line rocker bearing Cylindrical bearing Pot bearing[1] Spherical bearing[1]

Symbol	Function	Suitable type of bearing
◄─▯─►	Horizontal movements constrained perpendicular to rotational axis. Rotation about one axis only.	Roller bearing Constrained line rocker sliding bearing Constrained cylindrical sliding bearing Constrained pot sliding bearing[1] Constrained spherical sliding bearing[1]
▯ (with vertical arrows)	Horizontal movement constrained parallel to rotational axis. Rotation about one axis only.	Cylindrical sliding bearing Constrained line rocker sliding bearing Constrained pot sliding bearing[1] Constrained spherical sliding bearing[1]
◄─▯─► (with vertical arrows)	Horizontal movement in all directions. Rotation about one axis only.	Constrained roller sliding bearing Free rocker sliding bearing Free cylindrical sliding bearing Free pot sliding bearing[1] Free spherical sliding bearing[1]

Notes: All bearings can support vertical load unless otherwise indicated. Symbols represent plan view on bearing.
[1]Where there is no requirement to resist overturning movement normal to axis of rotation shown.

4.7 Positioning of bearings

Wherever possible, bearings should be placed on pedestals to protect them against water or dirt spilling from deck joints and against accumulations of dirt and debris likely to obstruct free movement. The stresses under rubber bearings are usually fairly low. However, if an elastomeric bearing is seated on a concrete plinth with a plan area approximately the same size as that of the bearing, potential tensile stresses developed during the movement of the bearing could cause disruption of the concrete around its edges. To ensure that the rubber of the bearing is adequately restrained and to avoid any tensile disruption of the concrete edge, the plan dimensions of the concrete should exceed those of the bearing by at least 50 mm, the actual dimensions depending upon the reinforcing detail. The steel reinforcement to the concrete should be designed to contain the designed stresses in a manner similar to that used for the design of prestressed concrete end blocks.

Bearings should normally be set level to avoid the effects of gravity. An exception to this is pairs of rubber bearings inclined towards each other to increase their shearing stiffness in one direction. Ideally bearings should be set horizontal in bridges having a cross-fall. The rotational axes of each bearing thus form an angle with the cross-sectional centre line which in practice corresponds to the cross-fall. Compression and displacements therefore occur when rotation takes place with elastomeric bearings. To ensure that these remain slight, stepping of the bearings corresponding to the cross-fall is recommended, and bearing dimensions should be kept to the minimum. If, for specific reasons, the bearings cannot be stepped, then allowances must be made for the resulting effects on the bearing, superstructure and substructure.

Special consideration must be given to elastomeric bearings supporting precast prestressed beams. Here the precamber of the beam can result in a considerable rotation at the ends and the bearings are generally unable to take up this rotation when under only the dead weight of the beam. Either tapering top plates have to be provided or the beam soffit recessed suitably, or else the beam must be bedded onto a stiff epoxy mortar placed on top of the bearing.

In the latter case the beam will have to be temporarily supported until the mortar has set. Under no circumstances should a pad be loaded on part of its top surface, as shown in Fig. 4.52(c).

As regards the positioning of elastomeric bearings the following precautions should be observed:

1. Bearings of different sizes must not be placed next to each other because of their different stiffnesses. By the same token, rigid bearings and rubber pads should not be placed along the same line of support unless the effect of the different stiffnesses is taken into account in the design of the substructure and superstructure.
2. Under no circumstances should two or more pads be placed one behind the other in the longitudinal direction of the bridge.

Setting tolerances for sliding bearings are very important as discussed in section 4.5.5.

Bearings should be located to ensure correct alignment beneath diaphragms or load bearing stiffeners. BS 5400: Part 9.2 gives a setting tolerance of ±3 mm. In the case of continuous girders, the mean level of bearings at any support shall be within ±0.0001 times the sum of adjacent spans, but not exceeding ±5 mm.

For all bridges, especially those with wide, skewed, horizontally curved or superelevated decks, due consideration should be given to the alignment of each bearing in regard to the actual direction of movement and rotation of the superstructure. In curved bridges and sometimes also in skew bridges, the movements resulting from temperature, shrinkage, creep and shortening due to prestressing run in fundamentally different directions. The use of roller bearings in such circumstances is not to be recommended.

The bearing guidance direction for curved bridges depends largely on the length of span. For short stiff spans, movement should be radial from the fixed point. For long flexible spans it can be along the line of the bridge. It is important to ensure that maintenance allows bridges to operate as designed.

Examples of the most suitable arrangement of bearings for certain typical situation are given in Fig. 4.53.

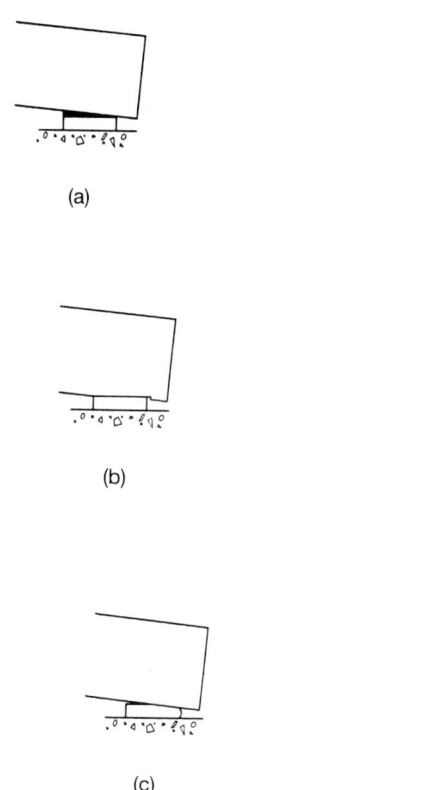

Fig. 4.52 Elastomeric bearing supporting precast beams. (a) Shaped bedding; (b) recessed beam; (c) partly loaded bearing.

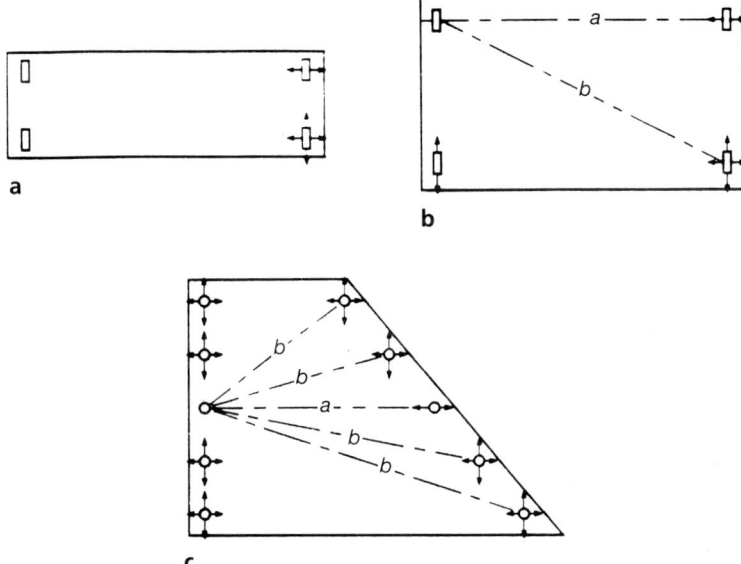

Fig. 4.53 Typical bearing arrangements. (a) For a narrow bridge: at the fixed support two fixed bearings with some play for movements transverse to the bridge axis and one longitudinally flexible and one multidirectional flexible bearing at the flexible support. To allow for manufacturing clearances within the fixed bearings they should both be capable of sustaining the full transverse load. (b) For very wide bridges, or if a fair degree of lateral movement must be allowed for, given special circumstances (e.g. prestressing of the end cross-member): one fixed and one transversely flexible bearing at the fixed support, one longitudinally flexible and one multidirectionally flexible bearing at the flexible support. (a = leading axis, b = direction of movement resulting from temperature and shrinkage.) (c) Slab bridge, one end extremely skewed. Since the tilt and roll axes for this type of deck usually run in different directions at the same bearing point, good bearing capacity can be achieved only by placing bearings which are flexible and tiltable in all directions with lateral restraint provided only by bearings on the bridge centreline. (a = leading axis, b = direction of movement resulting from temperature and shrinkage.)

4.8   Installation

4.8.1   GENERAL

Except in the case of elastomeric bearings, it is usual to fix bearings positively in position. The method of fixing bearings to the structure should ensure a totally secure connection, whilst allowing ease of removal of the bearings in the future without excessive jacking of the structure or other action which could result in overstress in the other bearings, discontinuity at the expansion joint, or other damage.

If bearings are to function efficiently it is important that they are installed properly. Bearings must be handled carefully and adequate lifting facilities provided when necessary. Temporary clamping devices must not be used for slinging or suspending bearings unless specifically designed for this purpose. Dust, dirt, grit, mortar or any foreign matter must not be allowed to enter the moving parts of bearings. For this reason, bearings should not be dismantled after leaving the manufacturer's works. If, for any reason it becomes necessary to do so, this should only be done under expert supervision and guidance from the manufacturer.

Distortion of bearings or damage of any sort should be prevented. During storage, bearings must be kept scrupulously clean and protected from excessive heat, sunlight, oils, fuels and other deleterious effects. Prior to fixing, bearings should be checked to ensure that they correspond with the data provided by the design engineer and any that show signs of deterioration, distortion or damage should be rejected as they are unlikely to function properly once installed.

The weight of the superstructure should not be transferred on to the bearings until sufficient strength has developed in the bedding material to resist the applied load. Temporary clamping devices should be removed at the appropriate time before the bearings are required to accommodate movement, and any holes exposed on the removal of temporary transit clamps filled. Where re-use of these fixing holes may be required, the material used to fill them should not only give protection against deterioration but also should be easily removable without damaging the threads.

Where necessary, suitable arrangements should be made to accommodate thermal movement and elastic deformation of the superstructure during construction. When provided, temporary supports under bearing baseplates should be compressible under design loading or removed once the bedding material has reached the required strength. Any voids left as a consequence of their removal should be made good using the same type of bedding material. Steel folding wedges and rubber pads are suitable for temporary supports under bearing baseplates.

Either provision has to be been made in the design of the bearings to permit them to be installed at any temperature likely to be met during construction (which should be the norm), or allowance made for the temperature prevailing at the time the bearings are installed. Presetting of bearings is fraught with difficulties and is to be avoided if at all possible.

4.8.2   BEDDING

It is essential to ensure that bearings are evenly supported over their entire area and that they are accurately aligned, particularly those with a single axis of rotation or horizontal restraint. Thus bedding surfaces must be flat and free from high or hard spots. Trowelling of concrete tends to produce a surface that is slightly rounded and it may be preferable to screed off with a straight edge. Whatever method is employed, an excessively smooth finish must be avoided as this will adversely affect the restraining friction of elastomeric bearings bedded directly onto the concrete substructure or the bond with any bedding mortar.

To overcome surface irregularities some form of bedding material must be employed. The choice of bedding material is influenced by the method of installing the bearings, the size of the gap to be filled, the strength required and the required setting time. When selecting the bedding material, consideration should therefore be given to the following factors:

1. type of bearing;
2. size of bearing;
3. loading on bearing;
4. construction sequence and timing;

5. early loading;
6. friction requirements;
7. dowelling arrangements;
8. access around the bearing;
9. thickness of material required;
10. design and condition of surfaces in the bearing area;
11. shrinkage of the bedding material.

It is essential that the composition and workability of the bedding material is specified with the above factors in mind. In some cases it may be necessary to carry out trials to ascertain the most suitable material particularly if flowable mortar is required. Commonly used materials are cementitious or chemical resin mortar, grout and dry packing. The use of materials such as lead, which tend to flow under load, leaving hard spots, should be avoided.

Epoxy mortar, usually only used in thin beds, allows little tolerance. Flowable grouts allow more tolerance for steel or precast concrete units, but can cause problems if they do not support bearings uniformly. Provided that it can be fully compacted, dry pack mortar is often the best choice. However, this becomes extremely difficult with large bearings.

The three main types of bedding material are based on cement, epoxy resin and polyester resin. Comparative properties are given in Table 4.12.

4.8.3 FIXING ELASTOMERIC BEARINGS

Elastomeric bearings may be laid directly on concrete, provided it is within the specified tolerance for position, level, flatness and smoothness and provided that the vertical force of the bearing is at least five times the horizontal force exerted by the bearing in resisting translational movement. For better control of these requirements it is usual to provide a thin layer of bedding material.

A number of fixing methods are available where location by friction alone is not possible. Bearings can be located in a slight recess provided that the outer reinforcing plates are not below the recessed surface, thereby preventing full shear movement. Dowels attached to the outer

bearing plates and set into recesses in the structure are sometimes used. The most common practice is to use a suitable adhesive. Epoxy mortars fulfil the dual purpose of bedding layer and adhesive. However, problems have been experienced with bearings slipping at the contact surfaces due to the mortar surfaces becoming contaminated with wax migrating from the rubber and inadequate curing of the adhesive.

4.8.4 FIXING OF BEARINGS OTHER THAN ELASTOMERIC

Bearings that are free sliding in all horizontal directions may not need to be positively fixed to the main structure if they are always subject to adequate vertical loading, since horizontal movement will occur on the plane of least resistance which is, of course, the bearing sliding surface. Nevertheless it is prudent to cater for vibration and accidental impact and some fixing should be provided. Shear keys or holding-down bolts should be accurately set into recesses cast into the structure using templates, and the remaining voids in the recesses should be filled with a material capable of withstanding the loads involved. Close tolerance bolts should be set using the bearings as templates. In this case special precautions should be taken to prevent contamination of the bearings during bolt installation.

Bearings that are to be temporarily supported should be firmly fixed to the substructure by the holding-down bolts or other means to prevent disturbance during subsequent operations, care being taken not to deform the bearing by overtightening the holding-down bolts. The gap between the substructure and the underside of the bearing is then filled with well-rammed dry pack mortar, grout or other suitable bedding material, the steel wedges or other hard packing removed and the resulting voids filled.

Alternatively bearings may be fixed directly to metal bedding plates cast in the supporting structure to the correct level and location or set on a bed of sand/cement mortar or a thin bed of synthetic resin mortar. Cementitious mortar beds should preferably be housed in recesses suitably reinforced on all sides.

It is possible to cast in the fixing bolts and position the bearing in one operation, in which case the bearing is supported on rubber

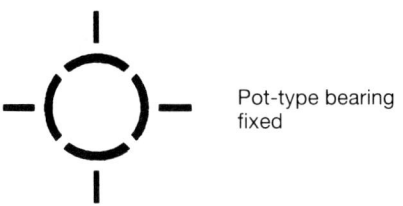

Pot-type bearing
fixed

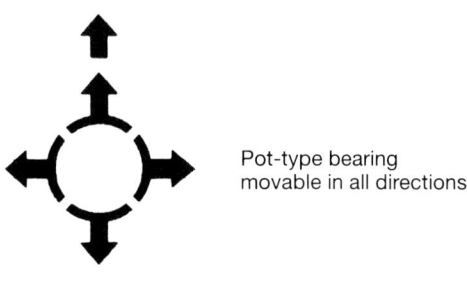

Pot-type bearing
movable in all directions

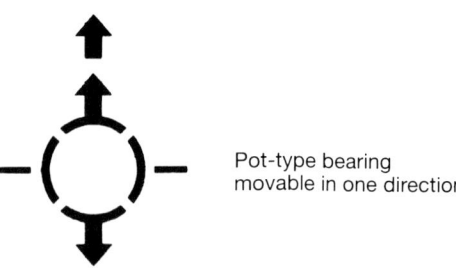

Pot-type bearing
movable in one direction

Fig. 4.55   Captions on top surface of bearings.

4.9   Testing

4.9.1   GENERAL

To maintain quality control and ensure that bearings perform satisfactorily throughout their life it is necessary to carry out tests both on the materials used and, in certain circumstances, on complete bearings. The various tests to which materials should be subject are enumerated in BS 5400: Part 9: Section 9.2 and are themselves covered by British or International Standards.

4.9.2   TESTING OF COMPLETE BEARINGS

The need for testing complete bearings depends upon the evidence available from previous tests and the level of quality control in the manufacture of the bearings. There are three types of bearing test: prototype, production and acceptance.

Prototype tests are used to verify the adequacy of bearing design and are carried out during the development stage. Production tests are performed to ensure that the correct materials and procedures have been used in the manufacture of bearings. Acceptance tests are used to confirm that complete bearings comply with the specified requirements.

In deciding on the need for and the type of acceptance testing the following factors should be taken into consideration:

1. The availability of existing well attested and relevant documented information.
2. The conditions under which the bearing will operate in service and the effect that these are likely to have on the bearing performance.
3. The relative importance of the bearing in the structure as a whole.
4. The complexity of the bearing.
5. The degree of quality control during manufacture.
6. The ease of inspection and replacement of the bearing after installation, should the need arise.
7. The availability of facilities for carrying out the required tests.

Bearing testing can be expensive, particularly for large bearings for which testing equipment is limited. Additional bearings need to be manufactured if tests are required to the ultimate limit stage. Where possible, conditions likely to adversely affect the bearings in service, e.g. extremes of temperature, variations in the rate of loading, rubber subject to sunlight, lack of lubrication, should be simulated in the tests.

4.9.3   LOAD TESTING

For most bearings, other than elastomeric, load tests are performed to check the adequacy of the design at both serviceability and ultimate limit states. Both vertical and, where appropriate, horizontal loads are applied to give the worst loading condition. Where there is likely to be load transfer due to tilting of the bearing this must be taken into account. Before recording deflection measurements, the bearing parts should be bedded in by applying a load representing the serviceability limit state. After testing, bearings should be dismantled and examined for any damage.

It should be possible to use bearings that satisfactorily pass the serviceability load test in structures, but bearings that have undergone a test to the ultimate limit state are unlikely to be suitable for further use.

As loads on elastomeric bearings are limited by long service considerations, the application of a single load, either the serviceability or ultimate limit state design load, cannot be expected to produce any visible damage. It is, however, normal practice to conduct a vertical load test on all elastomeric bearings after manufacture. Any defects will usually show as uneven deformation of the bearing. The deflection between one-third and full test load can be used to check the stiffness value. This test is, however, likely to overestimate the bearing stiffness and if the bearing compressive stiffness is critical to the design of the structure a separate test should be performed. Prior to any measurement being recorded, a vertical load equal to the serviceability load should be applied to the bearing and released. The vertical test load can then be applied in increments with the deflection recorded at each increment of load after any short-term creep has ceased. Similar measurements are taken as the load is released. The vertical deflection of the bearing between one-third and full test load for the last load cycle is used to calculate the vertical stiffness of the bearing.

4.9.4   TESTING RESISTANCE TO MOVEMENT

The design values for friction of PTFE running against stainless steel are based on long-term values which allow for running-in and loss of lubricant. Therefore friction tests on new bearings will not be representative of design values and usually give very low values of the coefficient of friction. Thus they serve little purpose. For details of long-term friction tests, refer to, e.g. [18–21].

Where well-attested test results are not available for the shear stiffness of elastomeric bearings, a shear stiffness test as described in Appendix A of BS 5400: Part 9.2 can be used. This consists essentially of loading a pair of identical bearings mounted in a compression testing machine with an intermediate steel plate deflected sideways an amount equal to the total thickness of elastomer in one bearing. The horizontal load/deflection curve is plotted and the shear stiffness can be obtained as the slope of the chord between $u = 0.05t_q$ and $u = 0.25t_q$ where u is the horizontal deflection and t_q is the total thickness of the elastomer in the bearing in shear as measured in the unloaded state.

Refere

1. Briti
   Com
2. LEE,
   Join
3. LON
   Butt
4. Low
5. Mor
   Soci
6. Briti
   Prot
7. Briti
   at B
8. Lon
   Pain
9. Buc
   Insti
10. Tim
    York
11. Roa
    York
12. Briti
    Com
13. Bla
    Tran
14. Briti
    Mec
    Insp
    Carb
    Requ
    Part
15. Briti
    Com
16. Kau
    Dura
    Publ
17. Tayl
    Tran

e

f

5.2 Medway Bridge, Rochester, Kent, UK

Length   11 west approach spans totalling 411 m; three river spans totalling 343 m; seven east approach spans totalling 243 m

Horizontal alignment   Straight

Width   41.9 m overall (typical approach viaduct); 34.6 m overall (river spans)

Superstructure type   River spans: the anchor spans and cantilevers are dual independent cast *in-situ* prestressed concrete box members with a central suspended span of precast prestressed beam and *in-situ* slab construction similar to the approach spans

Approach spans: precast prestressed concrete I- and box-beams composite with an *in-situ* concrete slab

Substructure type   River piers: solid reinforced concrete from the foundation to 1.83 m above high water level and cellular construction above this with concrete hinges at the top

Approach span piers: cast *in-situ* reinforced concrete portal frames

Foundation type   Piles or spread footings on chalk depending on the level of suitable chalk below ground or water level

Bearings   River piers incorporate continuous concrete hinges at the top The suspended river span is hinged at one end and on Meehanite cast iron roller bearings at the other

Approach viaduct piers carry Meehanite cast iron roller bearings

Movement joints   Cantilever interlocking toothed plate (comb) joints in carriageways and sliding plates in cycle tracks, footpaths and verges

General arrangement and articulation (Fig. 5.2(a))   Anchored at abutments and the main river piers

Reference

KERENSKY, O.A. and LITTLE G. (1964) Medway Bridge design. *Proceedings of the Institution of Civil Engineers*, **29**, 19–52. (Paper No. 6799.)

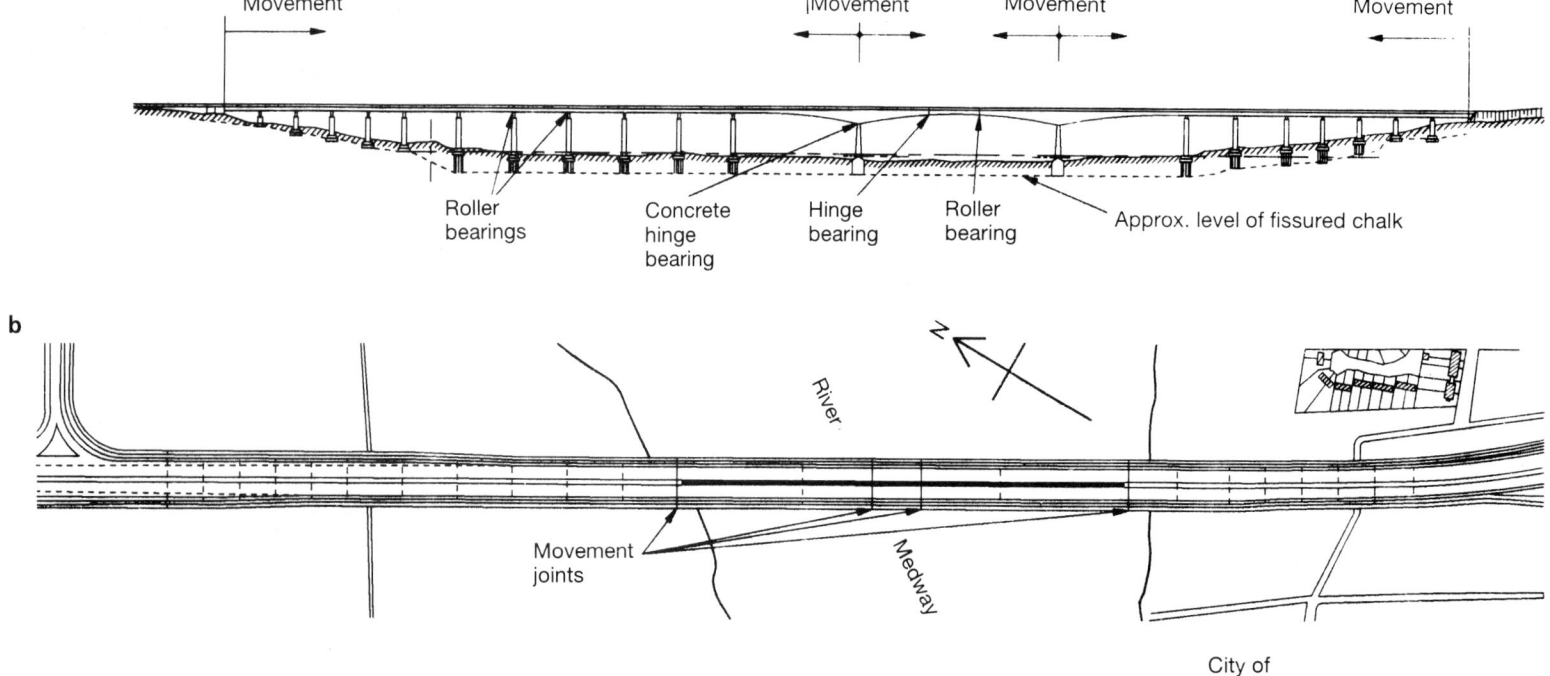

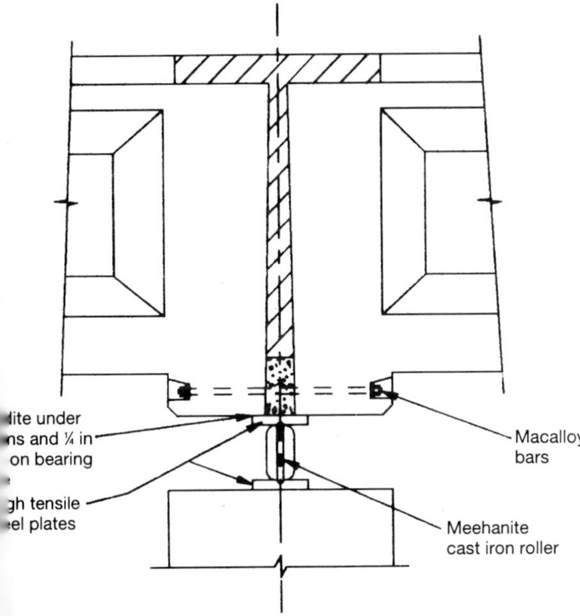

dite under
ns and ¼ in
on bearing

gh tensile
el plates

Macalloy
bars

Meehanite
cast iron roller

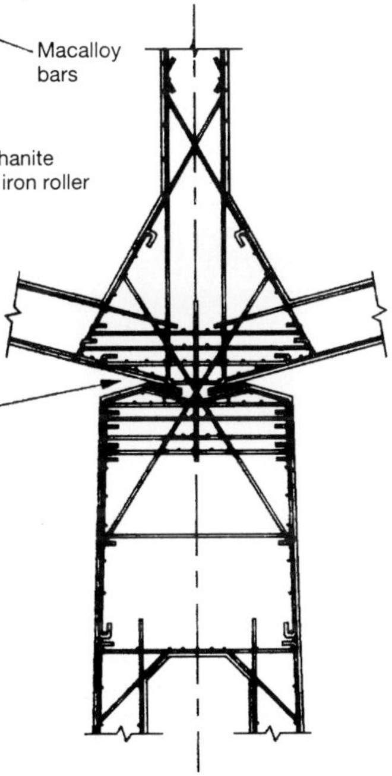

Sealer over
impregnated
obechi filler

d

e

Fig. 5.2 Medway Bridge, Rochester, Kent, UK, (a) articulation diagram.
(b) plan; (c) section through prestressed connection of viaduct beams;
(d) reinforced concrete hinge on main piers; (e) general view.

5.3   Mancunian Way elevated road, Manchester, UK

Length   32 spans totalling 985 m

Horizontal alignment   Multiple curves

Width   18.6 m overall (two-lane portion); 24 m overall (three-lane portion)

Superstructure type   Continuous prestressed concrete twin spine boxes with cantilevers

Substructure type   Solid rectangular reinforced concrete tapering to and monolithic with the pile cap

Foundation type   Bored cylinders 0.91 m and 1.37 m diameter

Bearings   Rotaflon pot sliding bearings in pairs on each column

Movement joints   Comb type joints at end abutments with stainless steel sliding plate joints in the median and footways

General arrangement and articulation (Fig. 5.3(a))   Anchored at Cambridge Street ramps

Reference

BINGHAM, T.G. and LEE, D.J.(1969) The Mancunian Way elevated road structure. *Proceedings of the Institution of Civil Engineers*, **42**, 459–92. (Paper No. 7186.)

b

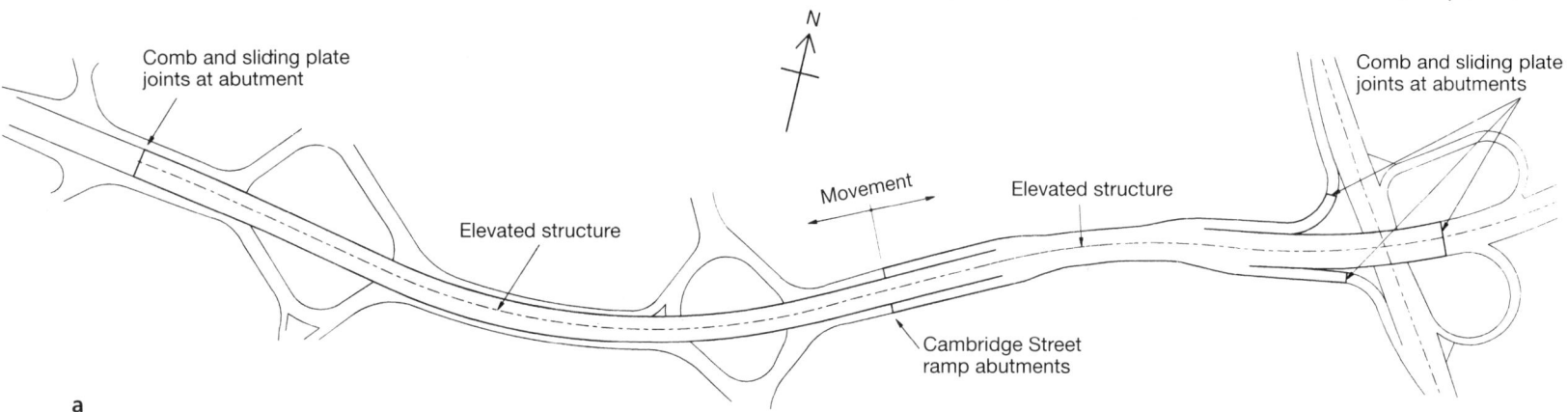

a

Comb and sliding plate joints at abutment

Comb and sliding plate joints at abutments

N

Elevated structure

Movement

Elevated structure

Cambridge Street ramp abutments

Fig. 5.3 Mancunian Way, Manchester, UK, (a) articulation diagram; (b) view during construction showing Rotaflon bearings in position at top of columns; (c) general view.

5.4   Western Avenue Extension (Westway), London, UK

	Section 1	Section 5	Section 6
Length	1189 m elevated through route with an elevated roundabout and slip roads	19 spans totalling 1158 m	38 spans totalling 975 m. Single and double deck structure with slip roads
Horizontal alignment	Through route straight at west and curving at east end	Multicurved	Multicurved
Width	26.2 m overall at west end 18.9 m overall above roundabout 28.7 m overall at east end	28.7 m overall	21.9 m overall upper deck 14.3 m overall lower deck
Super-structure type	Continuous precast prestressed concrete box girder with cantilevers	Continuous precast prestressed concrete box girder with cantilevers	Precast concrete hollow box beams with *in-situ* topping and diaphragms
Sub-structure type	*In-situ* concrete columns integral with pile cap	*In-situ* concrete columns integral with pile cap Reinforced concrete box anchorage abutment	Steel box columns integral with the pile cap at anchorage points Steel box portal frames composite with the superstructure at other positions
Foundation type	1.07 m diameter bored cylinders	1.37 m diameter bored cylinders below columns and 1.52 m diameter bored cylinders below anchorage abutment	0.91 m diameter bored cylinders

	Section 1	Section 5	Section 6
Bearings	Rotaflon pot sliding bearings	Glacier anti-clastic sliding bearings on columns and (vertical) laminated rubber bearings in prestressed joint at anchorage abutment	Steel spherical bearings at tops of columns at anchorage points. Metalastik guided sliding bearings below portal frame legs
Movement joints	Comb type joints at both ends of the through route and at slip road abutments	Demag rolling leaf joint at west abutment	Comb type joints at midlength and at abutments

Note: Sections 2 and 3 are at ground level and section 4 consists of simply supported spans on elastomeric bearings with buried expansion joints at each side of crosswall type supports.

	Section 1	Section 5	Section 6
General arrange-ment and articulation (Fig. 5.4)	The point of stagnant movement is located near the centre of the elevated roundabout	Anchored at east abutment	The two continuous 19 span lengths are both restrained by their midlength span

References

1. BAXTER, J.W. LEE, D.J. and HUMPHRIES, E.W. (1972) Design of Western Avenue extension (Westway). *Proceedings of the Institution of Civil Engineers*, **51**, 177–218. (Paper No. 7469.)
2. LEE, D.J. (1970) Western Avenue extension: the design of section five. *The Structural Engineer*, **48**(3), 109–20.

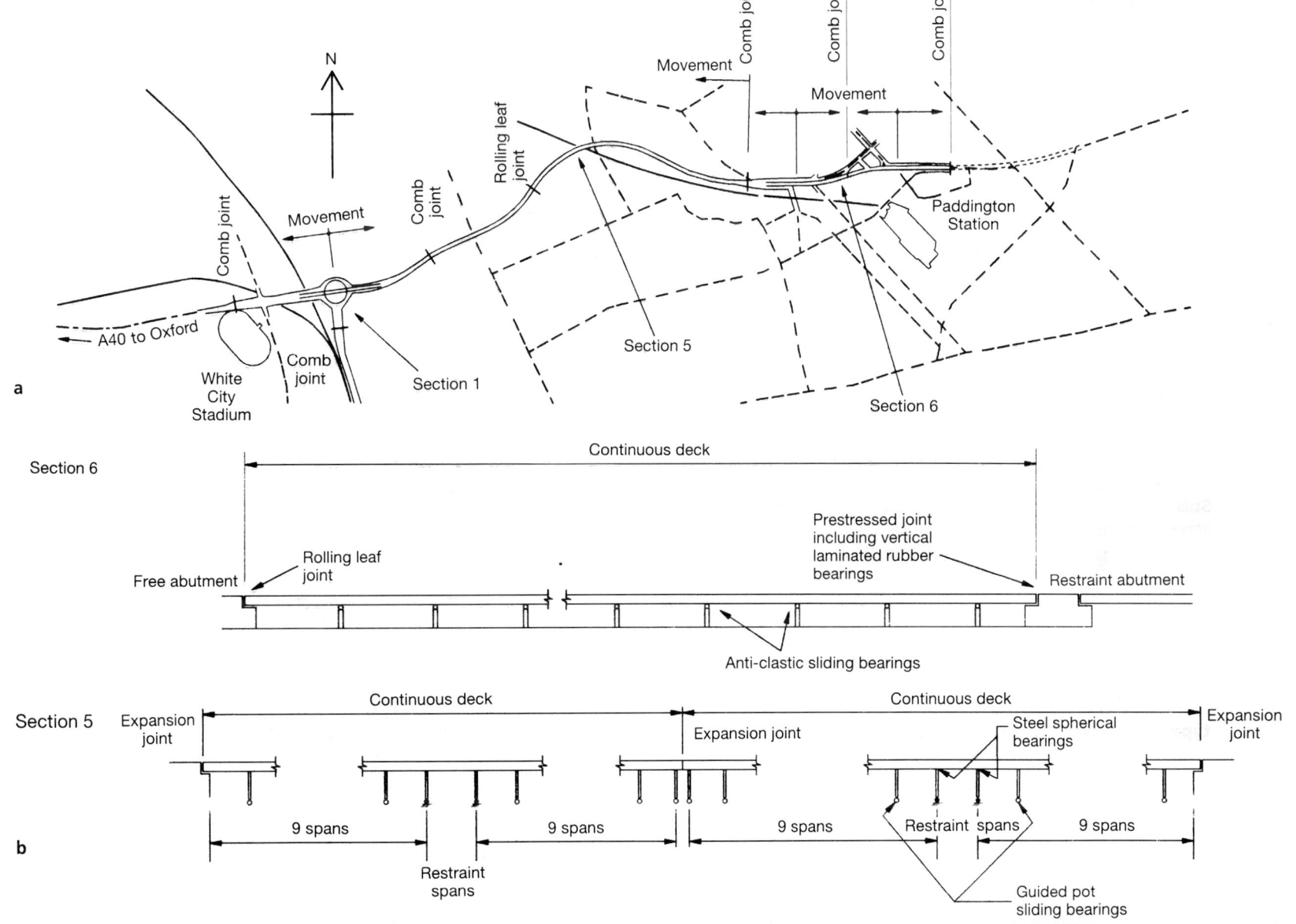

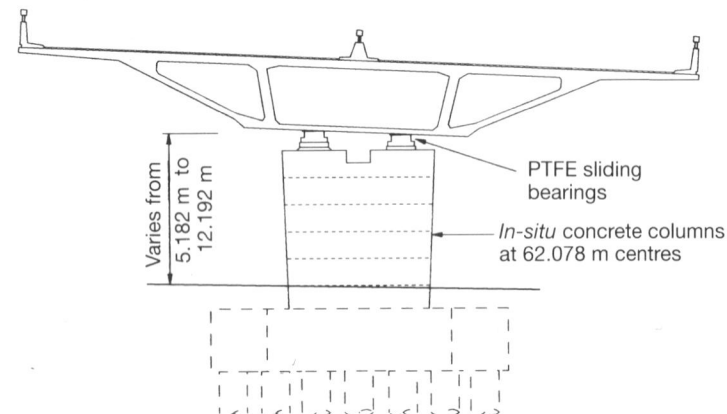

c

d

Fig. 5.4 Western Avenue Extension (Westway), London, UK, (a) plan; (b) articulation of Sections 5 and 6; (c) section 5: cross-section; (d) Rotaflon bearing at top of pier on Section 1 with flexible cover on bearing. Note use of X keepers before installation; (e) Demag expansion joints providing for a longitudinal movement of up to 826 mm; (f) general view during construction.

e

5.5 London Bridge, UK

Length Three spans totalling 262 m
Horizontal alignment Straight
Width 32.6 m overall
Superstructure type Cantilever and suspended span formed from four precast prestressed concrete boxes connected together transversely by *in-situ* strips
Substructure type Granite-faced reinforced concrete wall river piers
Foundation type Four circular steel-lined concrete shafts increasing in diameter in steps down to an under-reamed base in clay at both river piers
Bearings At the abutments and the ends of the suspended span PTFE/stainless steel sliding bearings with either cylindrical or spherical surfaces for rotation (PSC Tetron type) are used

Freyssinet hinges are used between the *in-situ* concrete pier diaphragm and the top of the pier
Movement joints Cast stainless steel comb joints, hooded at the suspended span joint and at the abutments where the maximum movement occurs
General arrangement and articulation (Fig. 5.5(a)) Anchored at river piers

References

1. Brown, C.D. (1973) London Bridge: planning, design and supervision. *Proceedings of the Institution of Civil Engineers*, **54**, 25–46. (Paper No. 7595.)
2. Brown, C.D. and Mead, P.F. (1973) London Bridge: demolition and construction. *Proceedings of the Institution of Civil Engineers*, **54**, 47–69. (Paper No. 7597.)

Fig. 5.5 London Bridge, UK, (a) articulation diagram; (b) plan; (c) general view. (Courtesy Mott MacDonald Ltd.)

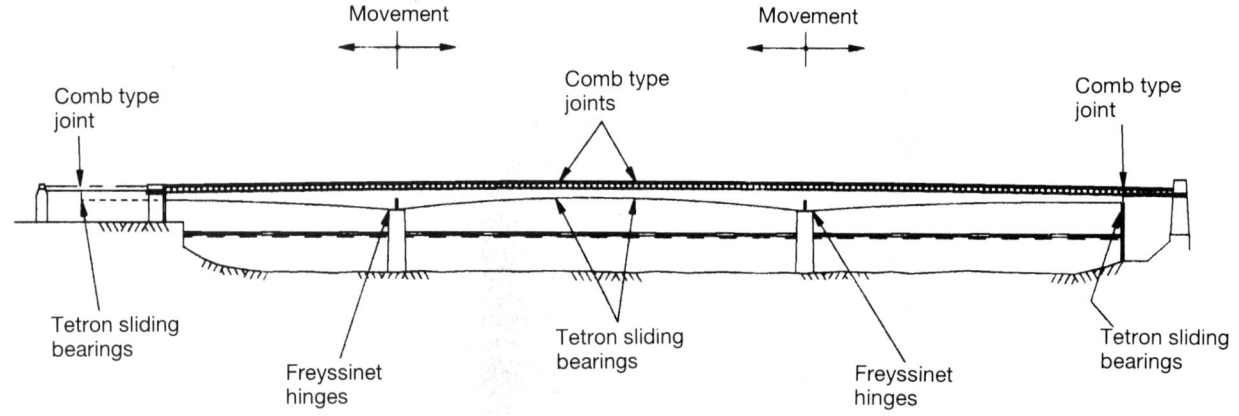

Movement

Movement

Comb type joint

Comb type joints

Comb type joint

Tetron sliding bearings

Freyssinet hinges

Tetron sliding bearings

Freyssinet hinges

Tetron sliding bearings

a

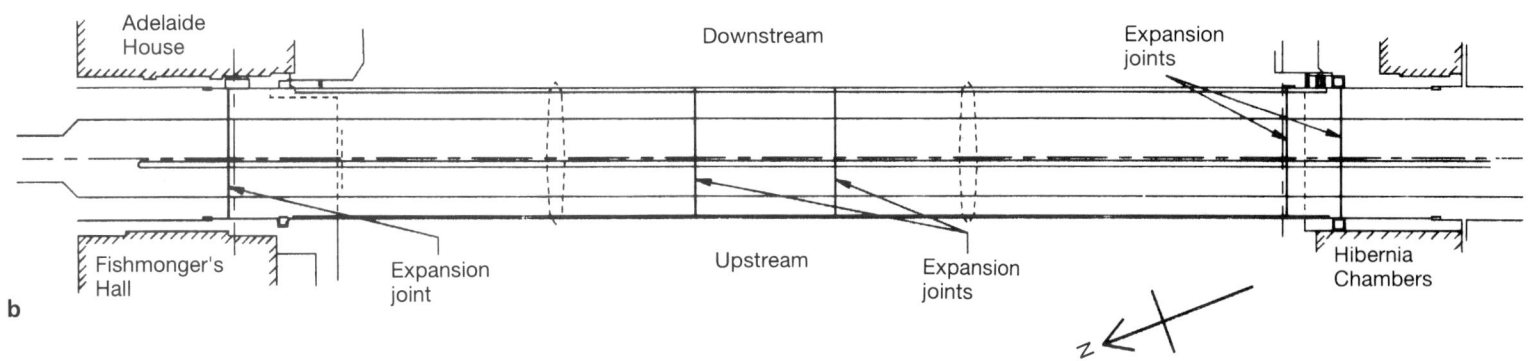

Adelaide House

Downstream

Expansion joints

Fishmonger's Hall

Expansion joint

Upstream

Expansion joints

Hibernia Chambers

b

5.6   Tyne and Wear Metro, Bridge N106, Newcastle/Gateshead, UK

Length   Three spans totalling 352.5 m

Horizontal alignment   Straight

Width   10.31 m overall

Superstructure type   Continuous steel through-truss with track ballast contained in longitudinal steel troughs supported by stringers spanning between cross girders at node points of the trusses

Substructure type   Twin reinforced concrete column bents

Foundation type   Rock anchored south abutment

    Mass concrete south river pier

    Steel H pile north river pier

    Bored pile north abutment

Bearings   Two hemispherical bearings at the south abutment

    Two sliding bearings at the south river pier

    Two stainless steel rollers on the north river pier

    Two hemispherical bearings on PTFE sliding surfaces at the north abutment

Movement joints   Scarfed rail joint in the continuous welded rails at the north abutment with a standard British Rail breather switch on the landward side of the abutment

General arrangement and articulation (Fig. 5.6(a))   Anchored at the south abutment

Reference

LAYFIELD, P., TAYLOR, G., MCILROY, P., KING, C. and CASEBOURNE, H. (1979) Tyne and Wear Metro: Bridge N106 over the River Tyne. *Proceedings of the Institution of Civil Engineers*, **66**, 169–89. (Paper No. 8205.)

Fig. 5.6   Tyne and Wear Metro, Bridge N106, Newcastle/Gateshead, UK. Articulation diagram.

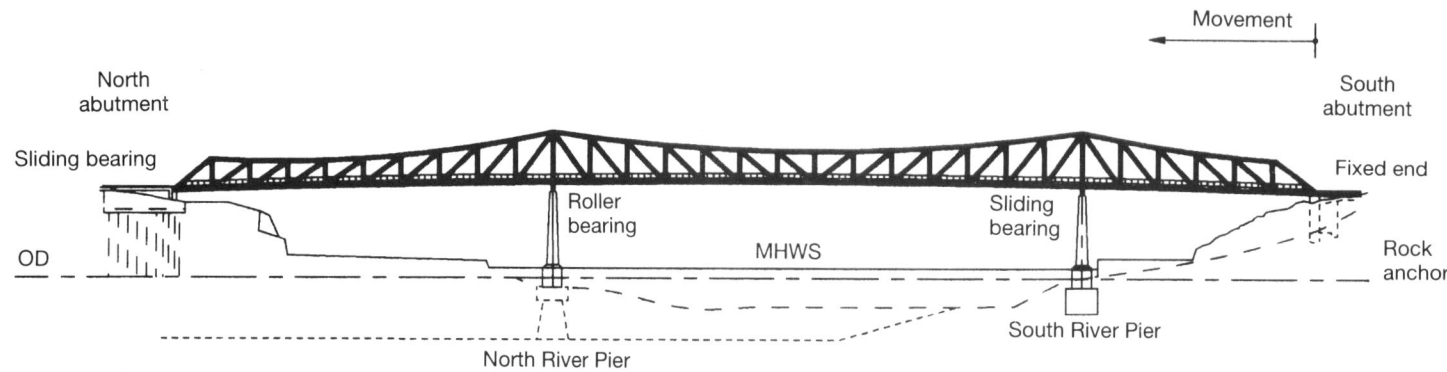

5.7   Orwell Bridge, Ipswich, Suffolk, UK

Length   Seven west approach spans totalling 413 m; three river spans totalling 402 m; eight east approach spans totalling 472 m.

Horizontal alignment   Straight except at ends of approach spans

Width   19 m overall

Superstructure type   Twin cast *in-situ* prestressed concrete box girders

Substructure type   Twin reinforced concrete columns of constant cross-section cantilevering up from a common pile cap support the approach spans, and single piers cellular over the middle third of their height carry the river spans

Foundation type   Bored piles 1.05 m diameter

Bearings   Two bearings beneath each box as follows: line rocker bearings on pier east of the navigation span: multiple roller bearings with a rocker on pier west of the navigation span; single roller bearings at anchor span piers and three adjacent piers to the east

    Lubricated PTFE/stainless steel sliding rocker bearings on remaining piers

Movement joints   At both abutments (Mageba)

Note   Extensive testing of movement bearings was carried out during fabrication

General arrangement and articulation (Fig. 5.7)   Anchored at pier on east side of navigation span

References

1.   LEWIS, C.D., ROBERTSON, A.I. and FLETCHER, M.S. (1983) Orwell Bridge – design. *Proceedings of the Institution of Civil Engineers*, **74**, 765–78. (Paper No. 8729.)

2.   VAN LOENEN, J. and TELFORD, S. (1983) Orwell Bridge – construction. *Proceedings of the Institution of Civil Engineers*, **74**, 779–804. (Paper No. 8730.)

3.   BUECHLER, W.R. (1981) *Proof Testing of Extremely High Loaded Bridge Bearings*. American Concrete Institute Special Publication, SP70, Detroit

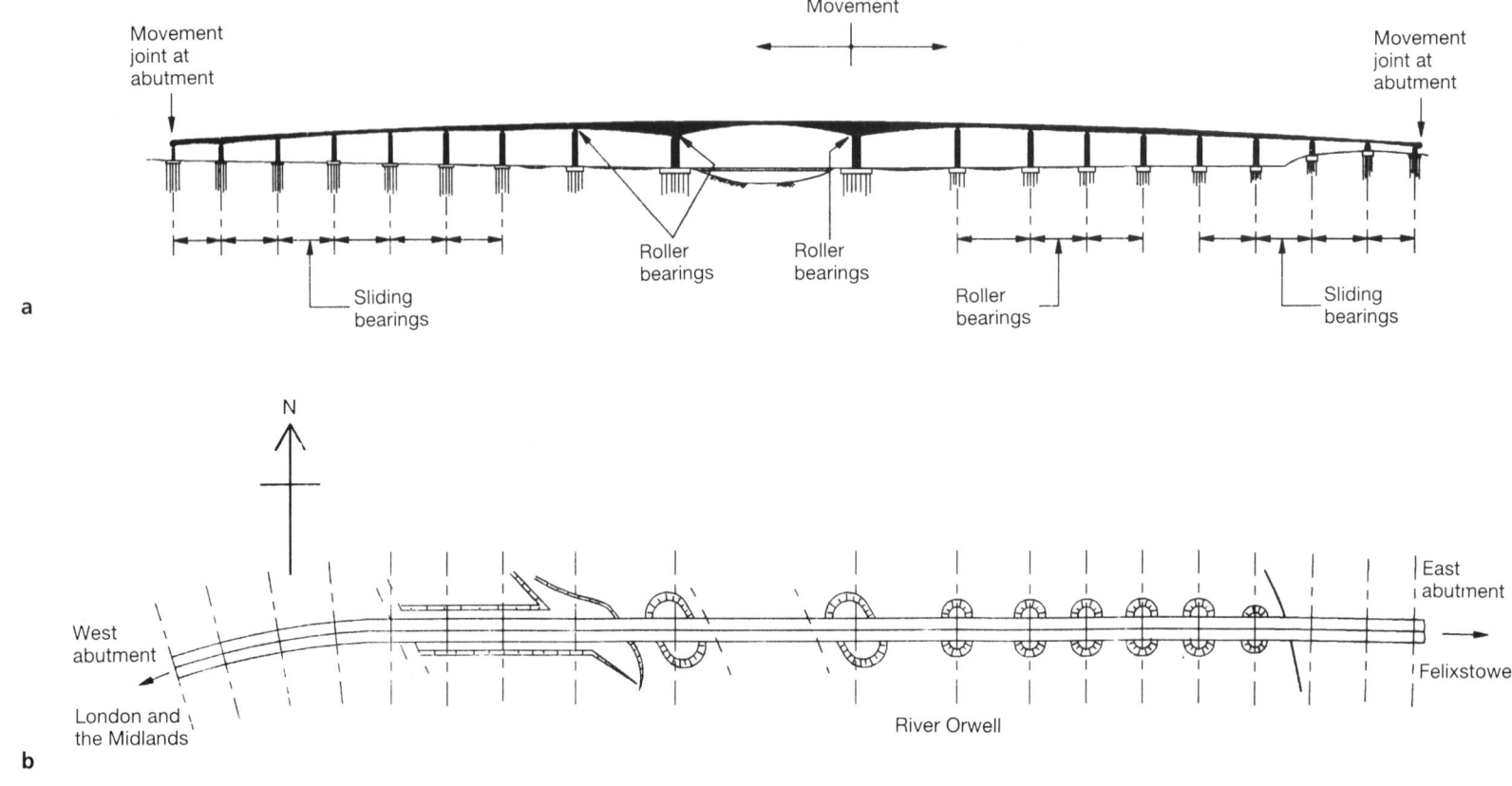

Fig. 5.7   Orwell Bridge, Ipswich, UK, (a) articulation diagram; (b) plan; (c) general view

5.8   Redheugh Bridge, Newcastle/Gateshead, UK

Length   Three main spans totalling 360 m
Horizontal alignment   Straight
Width   15.8 m overall
Superstructure type   Continuous prestressed concrete box girder cast *in-situ*
Substructure type   Reinforced concrete solid wall piers
Foundation type   Circular reinforced concrete caissions
Bearings   Freyssinet hinges between the superstructure and the top of the river piers. PTFE pot sliding bearings at the side piers
Movement joints   Prefabricated reinforced elastomeric units over each side pier
Note   Flexible river piers
General arrangement and articulation (Fig. 5.8(a))   Anchored at river piers

Reference
LORD, J.E.D., GILL, J. M. and MURRAY, J. (1984) The new Redheugh Bridge. *Proceedings of the Institution of Civil Engineers*, **76**, 497–521. (Paper No. 8750.)

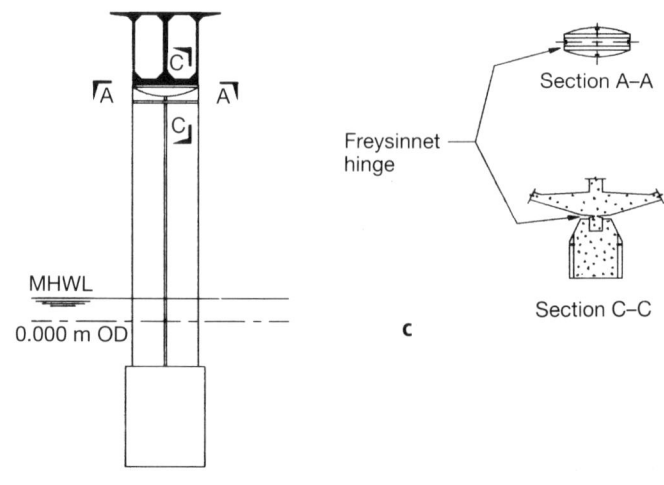

Fig. 5.8   New Redheugh Bridge, Newcastle/Gateshead, UK, (a) articulation diagram; (b) elevation on Pier N1; (c) Sections A–A and C–C through hinge; (d) general view. (Courtesy Mott MacDonald Ltd.)

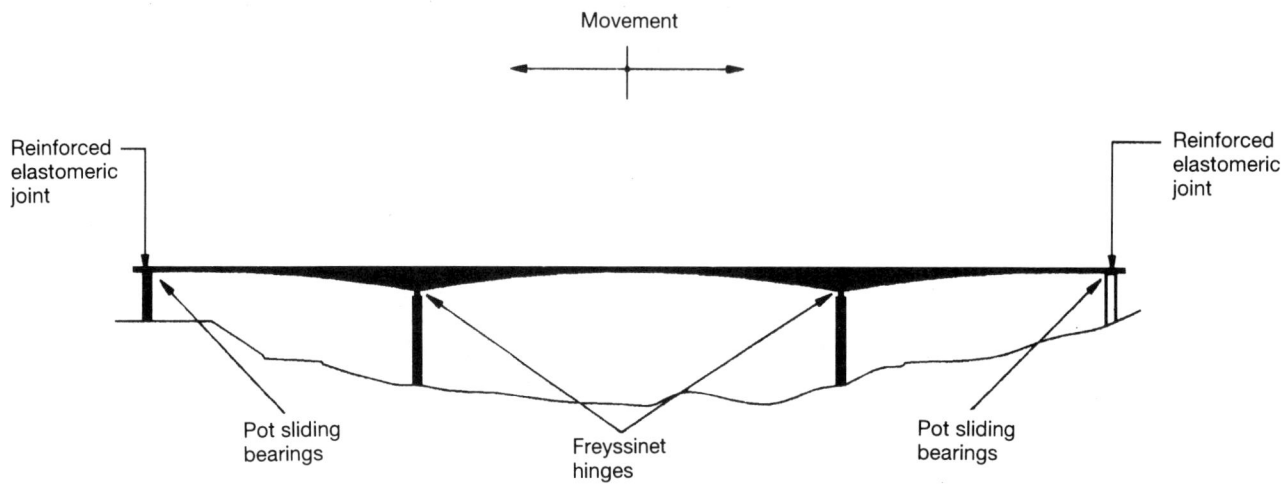

5.9 Foyle Bridge, Northern Ireland

Length   Three main spans totalling 522 m; seven east approach viaduct spans totalling 344 m

Horizontal alignment   Straight main spans and curved east approach spans

Width   Twin structures, with 100 mm median slot, totalling 22.6 m overall

Superstructure type   Main spans: continuous steel box girder with orthotropic steel plate deck

   Approach viaduct spans: continuous prestressed concrete cast *in-situ* box girder

Substructure type   Twin rectangular reinforced concrete piers. A steel box portal beam connects between the tops of the two river piers only

Foundation type   Spread footings on rock

Bearings   Main spans: tie back bearing at west abutment
   Fixed knuckle bearings on the river pier Hi-load roller bearings at east side pier
   Approach viaduct spans: restraint pot bearings at pier near midlength. One free and one guided sliding bearing at all other supports

Movement joints   Maurer D75 type joint at west abutment
   Maurer D720 type joint at east side pier
   Maurer D180 type joint at east abutment

Note   Flexible river piers

General arrangement and articulation (Fig. 5.9(a))   Anchored at west abutment and approximately at midlength of the approach viaduct.

References

1. WEX, B.P. GILLESPIE, N.M. and KINSELLA, J. (1984) Foyle Bridge: design and tender in a design and build competition. *Proceedings of the Institution of Civil Engineers*, **76**, 363–86. (Paper No. 8793.)
2. QUINN, W.N. (1984) Foyle Bridge: construction of foundations and viaduct. *Proceedings of the Institution of Civil Engineers*, **76**, 387–409. (Paper No. 8794.)

Fig. 5.9   Foyle Bridge, Northern Ireland, (a) articulation diagram; (b) plan; (c) main bearings at Piers 1 and 2; (d) Piers 1 and 2; (e) expansion joint at Pier 3; (f) girder tie-back at west abutment; (g) tie-back bearing at west abutment. See Fig. 1.4 for general view.

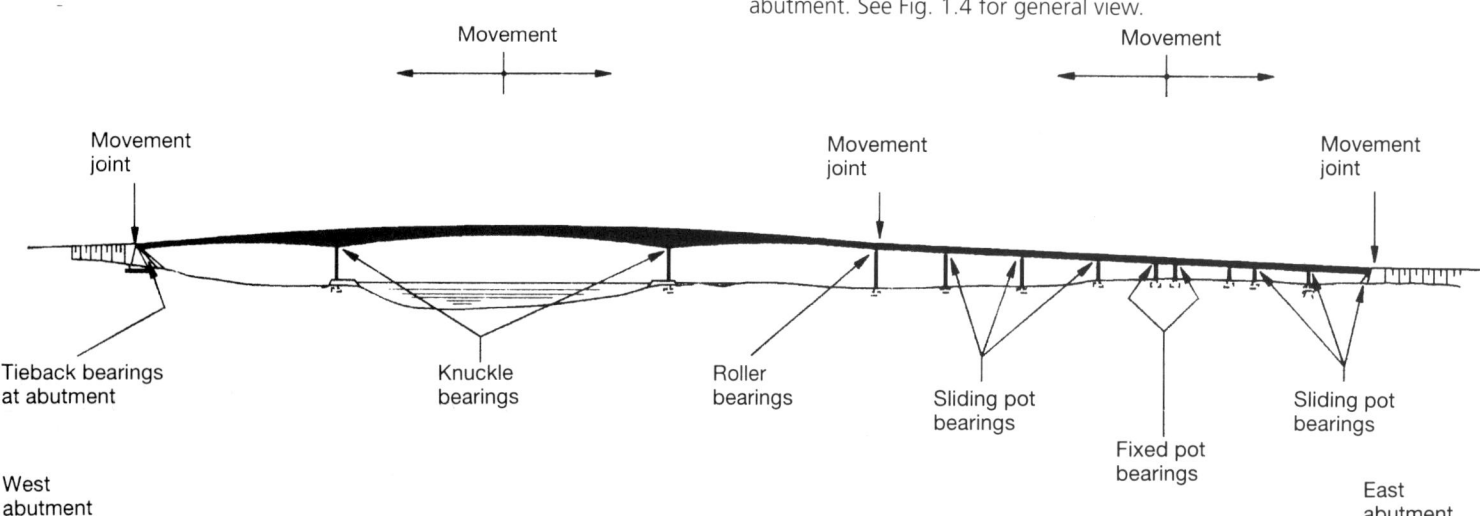

Movement

Movement

Movement joint

Movement joint

Movement joint

Tieback bearings at abutment

Knuckle bearings

Roller bearings

Sliding pot bearings

Fixed pot bearings

Sliding pot bearings

a West abutment

East abutment

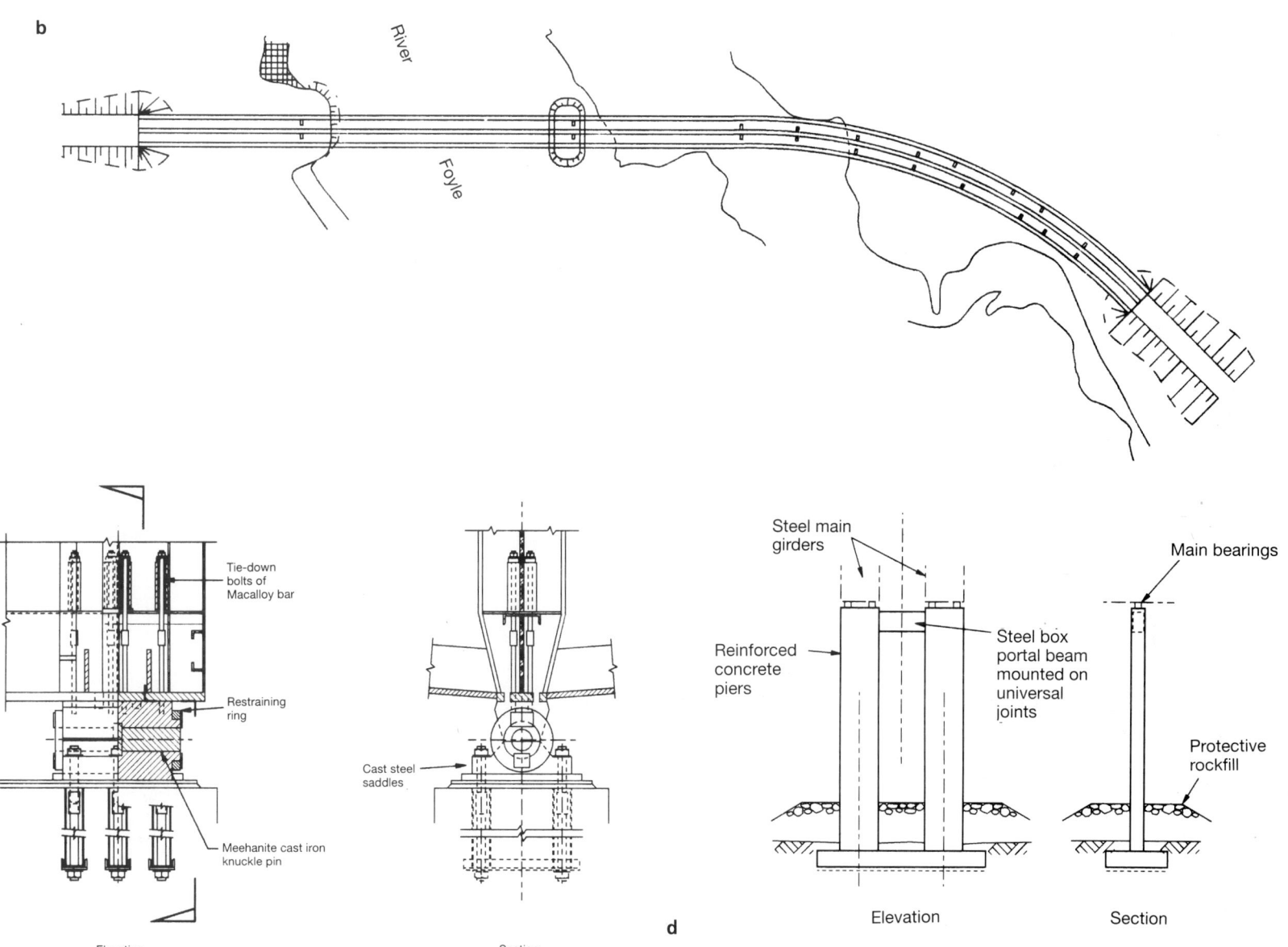

b

River

Foyle

c

Tie-down bolts of Macalloy bar

Restraining ring

Meehanite cast iron knuckle pin

Elevation

Cast steel saddles

Section

d

Steel main girders

Main bearings

Reinforced concrete piers

Steel box portal beam mounted on universal joints

Protective rockfill

Elevation

Section

e

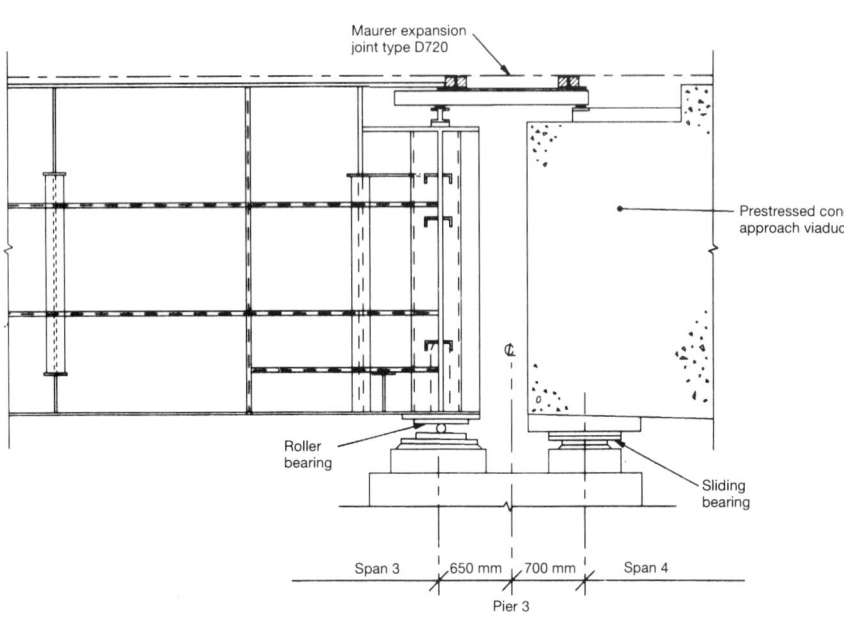

Maurer expansion joint type D720

Prestressed concrete approach viaduct

Roller bearing

Sliding bearing

Span 3 | 650 mm | 700 mm | Span 4

Pier 3

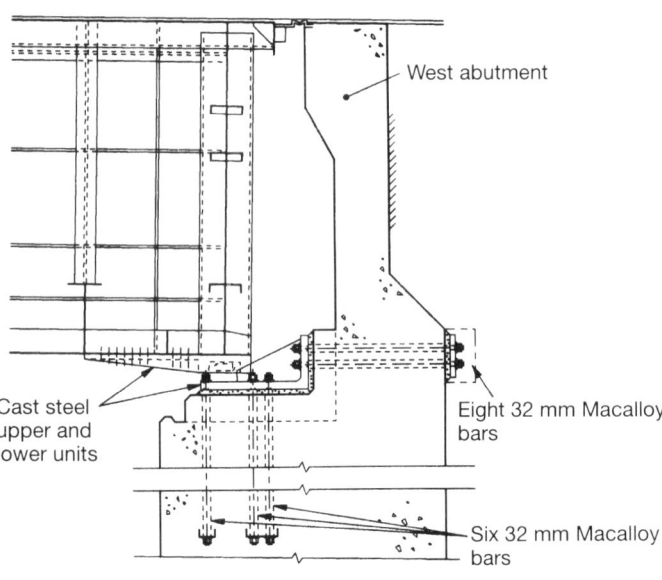

West abutment

Cast steel upper and lower units

Eight 32 mm Macalloy bars

Six 32 mm Macalloy bars

f

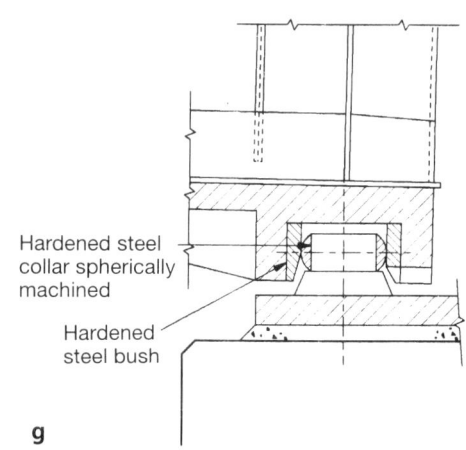

Hardened steel collar spherically machined

Hardened steel bush

g

5.10 Kylesku Bridge, Highland Region, Scotland

Length Five spans totalling 277.8 m
Horizontal alignment Curved
Width Varies from 8.4 to 9.28 m
Superstructure type *In-situ* and precast prestressed concrete box girder
Substructure type V-frame legs integral with the deck and base
Foundation type Spread footings on rock
Bearings Glacier spherical GY 350/300 and GX 400/300 at the abutments only

Movement joints Maurer Type D241 at both abutments
General arrangements and articulation (Fig. 5.10(a))
Anchored at V-frame legs

References
1. NISSEN, J., FALBE-HANSEN, K. and STEARS, H.S. (1985) The design of Kylesku Bridge. *The Structural Engineer*, **63A**(3), 69–76.
2. MARTIN, J.M. (1986) The construction of Kylesku Bridge. *Proceedings of the Institution of Civil Engineers* Part 1, **80**, 317–42.

Fig. 5.10 Kylesku Bridge, Highland Region, Scotland, (a) articulation diagram; (b) plan; (c) general view. (Courtesy Ove Arup & Partners.)

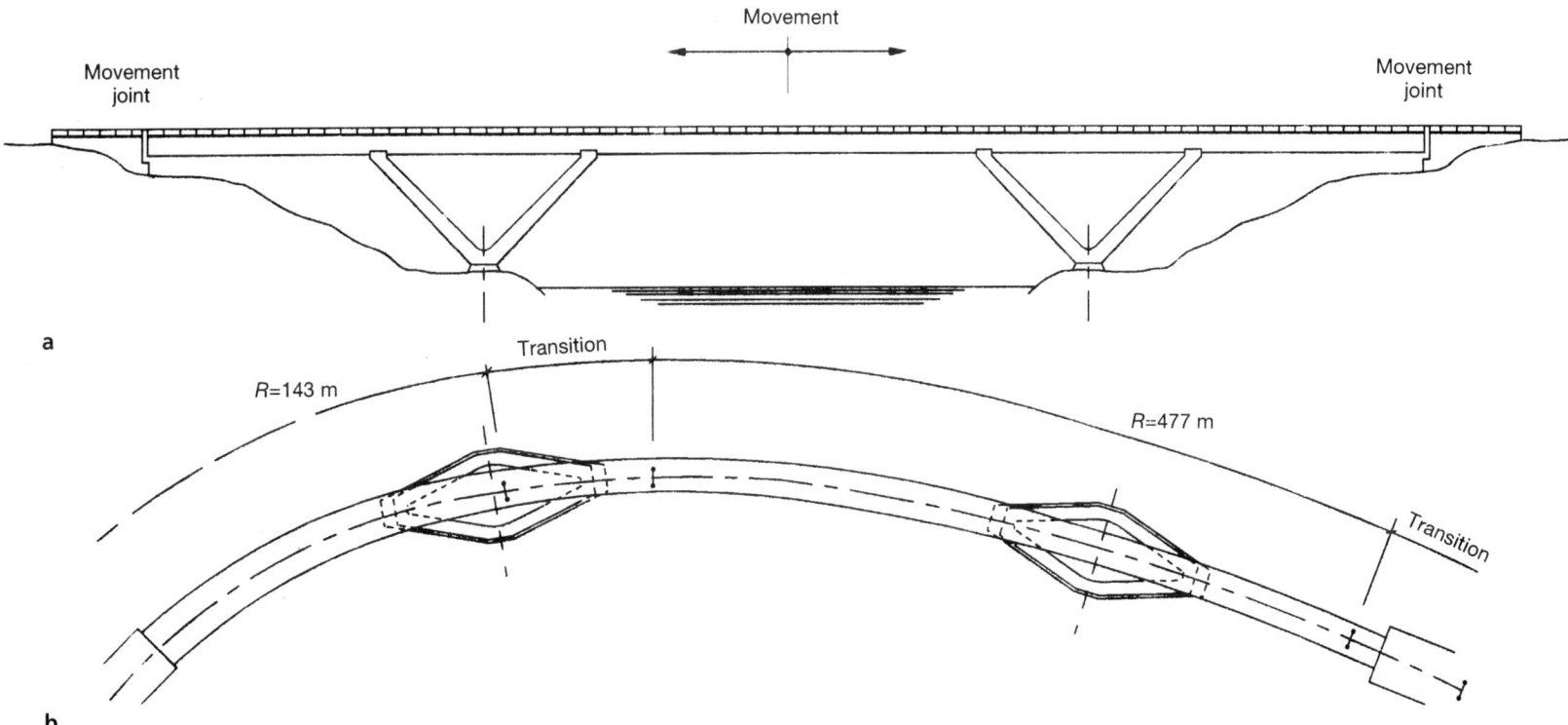

CAOLAS CUMHANN

5.11   Torridge Bridge, Bideford, Devon, UK

Length   Eight spans totalling 641.85 m

Horizontal alignment   Straight

Width   13.3 m overall

Superstructure type   Continuous precast prestressed concrete box girder with cantilevers

Substructure type   Solid reinforced concrete piers integral with the caisson and pile caps

Foundation type   Abutments and end piers: reinforced concrete spread footings on rock

    Intermediate piers: bored piles 2.1 m diameter on rock

    Main river piers: rectangular reinforced concrete caissons on rock

Bearings   Mageba spherical bearings with a separate sliding shear key which could be fixed to prevent longitudinal movement as required during construction

Movement joints   Mageba type LR7-A63 at both abutments

General arrangement and articulation (Fig. 5.11)   Anchored at the central river pier

Reference

POTHECARY, C.H. and CHRISTIE, T.J.C. (1990) Torridge Bridge design. *Proceedings of the Institution of Civil Engineers*, **88**, 191–260. (Paper No. 9424.)

Fig. 5.11   Torridge Bridge, Bideford, Devon, UK, (a) articulation diagram; (b) general view. (Courtesy MRM Partnership.)

a

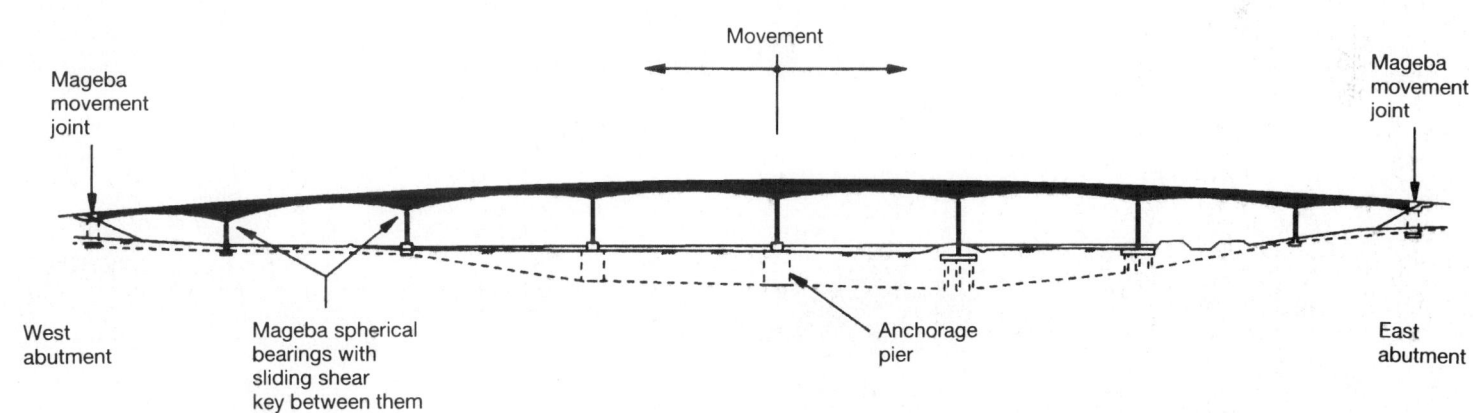

5.12   River Dee Viaduct, Newbridge, Clwyd, Wales

Length   Five spans totalling 348 m

Horizontal alignment   Straight

Width   12.3 m overall

Superstructure type   Continuous cast *in-situ* prestressed concrete box girder with cantilevers

Substructure type   Hollow rectangular reinforced concrete piers with fluted end faces

Foundation type   Spread footings except for the south abutment which is on 1.05 m diameter bored piles

Bearings   Two sliding pot bearings at the side piers, one guided and one free sliding

Two free sliding bearings at both abutments

Movement joints   At both abutments

Note   The two central piers are integral with the deck forming a portal frame

A differential settlement of 25 mm between the abutments and adjacent piers was allowed for

General arrangement and articulation (Fig. 5.12)   Anchored by central portal frame span

References

1. SAUNDERS, J.R. (1991) River Dee Viaduct – design. *Proceedings of the Institution of Civil Engineers*, **90**, 237–57. (Paper No. 9720.)
2. THOMAS, C.D. and BOURNE, S.D. (1991) River Dee Viaduct – construction. *Proceedings of the Institution of Civil Engineers*, **90**, 259–80. (Paper No. 9721.)

a

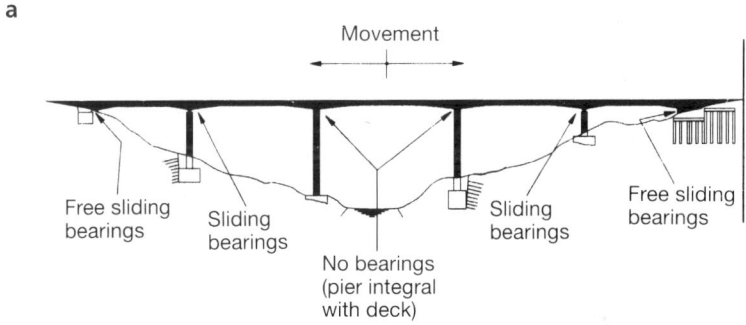

b

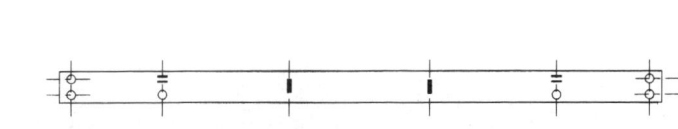

Fig. 5.12   River Dee Viaduct, Newbridge, Clwyd, Wales, (a) articulation diagram; (b) plan; O, horizontal sliding in any direction; =, guided; (c) River Dee Viaduct under construction. (Courtesy Travers Morgan Ltd.)

5.13 Narrows Bridge, Perth, Australia

Length Five spans totalling 335 m
Horizontal alignment Straight
Width 27.5 m overall
Superstructure type Continuous precast prestressed concrete I-beams connected by *in-situ* concrete strips
Substructure type Reinforced concrete piers hinged top and bottom
Foundation type 'Gambia' reinforced concrete driven piles 0.80 m diameter
Bearings Stainless steel rocker plates at top and bottom of piers for longitudinal movement and roller bearings in pile caps under outer columns for transverse movement. Stainless steel rocker bearings at the north abutment and roller bearings at the south abutment
Movement joints Rolling leaf carriageway joints and sliding plate footway joints at the south abutment
General arrangement and articulation (Fig. 5.13(a))
Anchored at the north abutment

Reference
BAXTER, J.W., BIRKETT, E.M. and GIFFORD, E.W.H. (1961) The Narrows Bridge, Perth, Western Australia. *Proceedings of the Institution of Civil Engineers*, **20**, 39–84. (Paper No. 8498.)

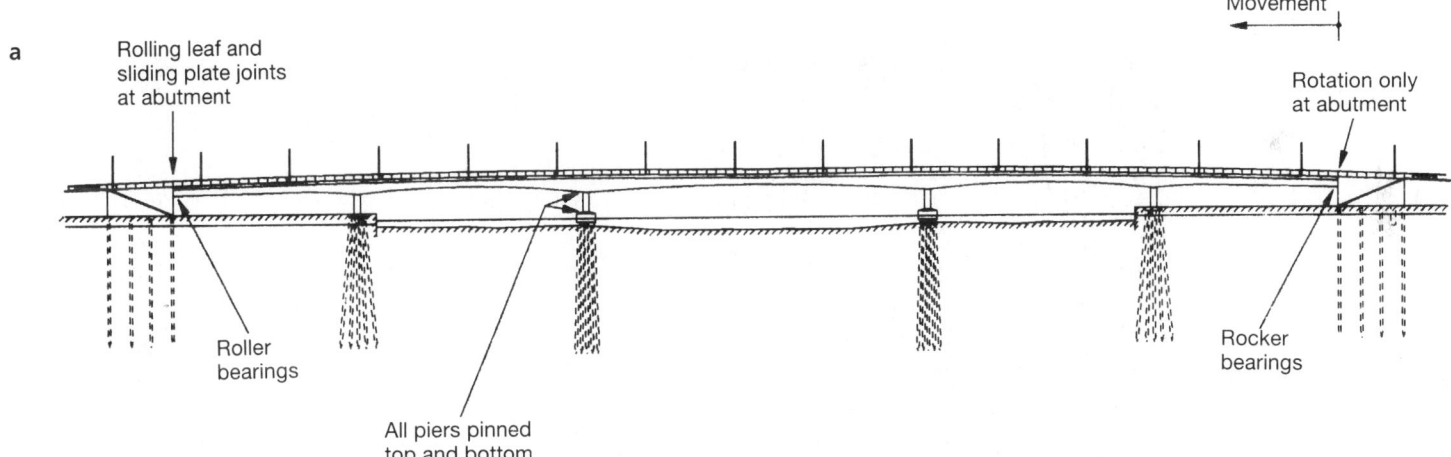

a

Rolling leaf and sliding plate joints at abutment

Movement

Rotation only at abutment

Roller bearings

All piers pinned top and bottom

Rocker bearings

Fig. 5.13 Narrows Bridge, Perth, Western Australia, (a) articulation diagram; (b) plan; (c) expansion joint; (d) general view.

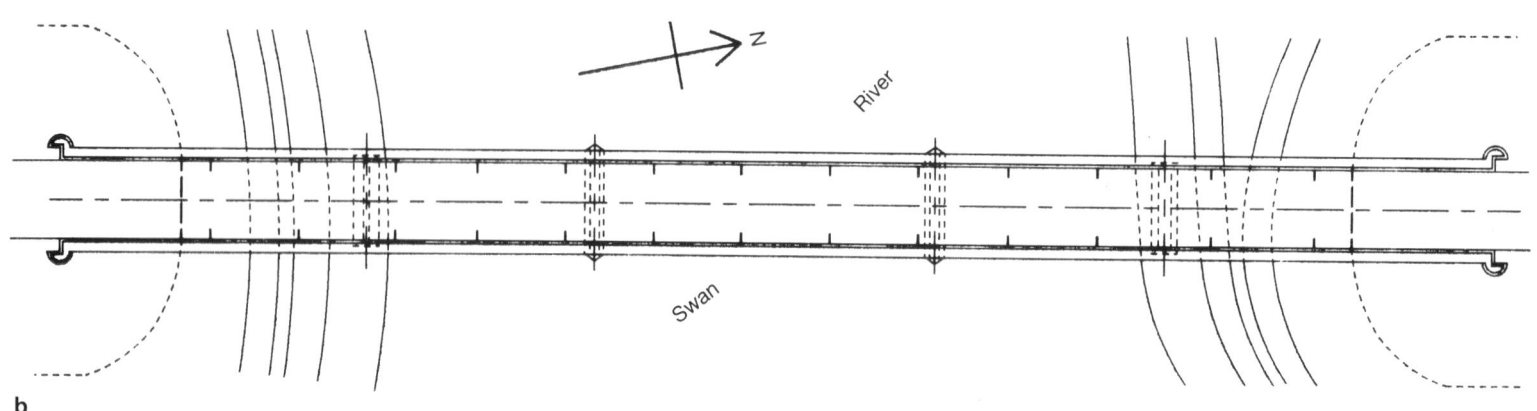

b

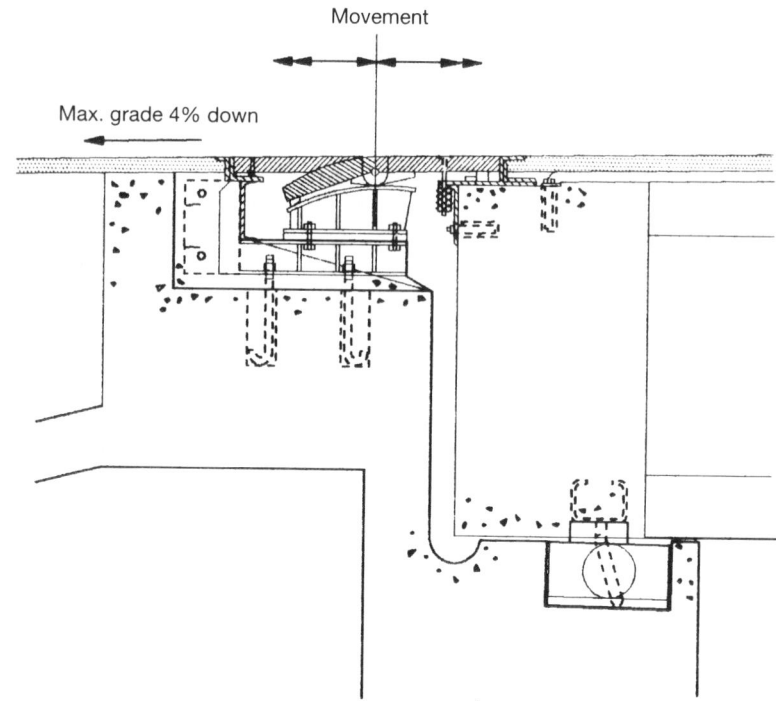

c

5.14 Westgate Bridge Approaches, Melbourne, Australia

Length   10 west approach spans totalling 655 m; 13 east approach spans totalling 855 m

Horizontal alignment   Curved

Width   37.34 m overall

Superstructure type   Continuous precast prestressed concrete spine box girder with cantilevered outer decks

Substructure type   Hollow reinforced concrete piers up to 43 m high hinged top and bottom

Foundation type   Bored cylinders 1.52 m diameter

Bearings   Hi-load rocker bearings top and bottom of piers

Movement joints   Demag rolling leaf joints between approach and main spans

General arrangement and articulation (Fig. 5.14(a))
Anchored at abutments

References

1. FERNIE G.N., WILSON, C.A. and KELLY, B. (1974) Westgate Bridge – design and construction of foundations, in *Annual Conference, Institution of Engineers* (Australia), Newcastle, NSW, pp. 219–25.
2. CRESSWELL, R., FERNIE, G.N. and LEE, D.J. (1974) Westgate Bridge – concrete approach viaducts – design and construction of substructure and superstructure, in *Annual Conference, Institution of Engineers* (Australia), Newcastle, NSW, pp. 227–34.

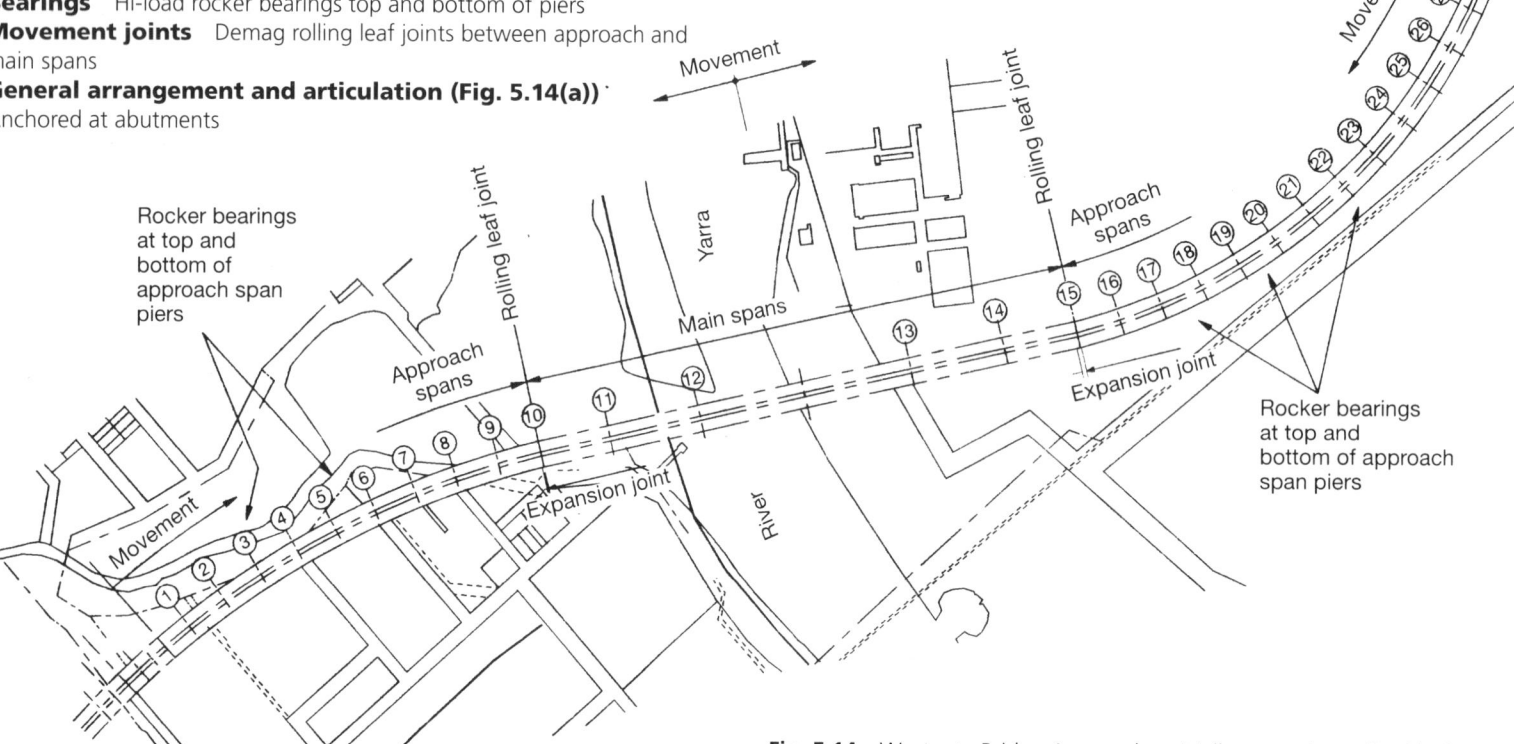

Fig. 5.14 Westgate Bridge Approaches, Melbourne, Australia, (a) plan; (b) general view.

5.15   Commonwealth Avenue Bridge, Canberra, Australia

Length   Five spans totalling 335 m, maximum span 73.2 m
Horizontal alignment   Straight
Width   Two structures, each 11.6 m overall
Superstructure type   Continuous concrete multi-cell box girders
Substructure type   Reinforced concrete octagonal central piers
Foundation type   Bored cylinders 1.83 m diameter
Bearings   Two Kreutz hardened steel roller bearings at top of each pier and at the abutments
Movement joints   Comb type joints at abutments
Note   The design allowed the bridge to be anchored at either abutment. During the final stages of construction the whole bridge was jacked along to the optimum position on its bearings. At this stage the rolling friction was found to be only 0.2%, well below the design value of 3%. This indicated a high quality and geometric accuracy of construction and does not justify reduction of the design friction value on other projects
General arrangement and articulation (Fig. 5.15)   Anchored at one abutment

References

1.   BIRKETT, E.M. and FERNIE, G.N. (1964) Bridges in the Canberra Central Lakes Area. *Journal of Institution of Engineers* (*Australia*), **36** (7–8), 139–50.
2.   LEE, D.J. (1967) Design of bridges of precast segmental construction. Technical Paper No.10, Design of Prestressed Concrete Bridge Structures Symposium, Concrete Society, 6 June.

Fig. 5.15   Commonwealth Avenue Bridge, Canberra, Australia, (a) during construction, showing roller bearings at top of piers; (b) after completion; (c) Kreutz roller bearings on the piers.

c

5.16 Gladesville Bridge, Sydney, Australia

Length   305 m span arch with four approach spans each end. Overall length 578.5 m

Horizontal alignment   Straight

Width   26 m overall

Superstructure type   Continuous precast, prestressed concrete T-beam deck with infilled concrete strips. The supporting fixed ended arch comprises four separate rectangular hollow concrete ribs

Substructure type   Flexible prestressed concrete cantilever column bents with concrete hinge connections to the deck

Foundation type   Reinforced concrete spread footings in shallow excavations on sandstone

Bearings   Roller bearings at expansion joint halved joints and Freyssinet type hinges between deck and substructure

Movement joints   Cantilever interlocking toothed plate (comb) joints

Note   The articulation adopted takes advantage of the fixity at the abutments and in the arch itself and the deck movements at each expansion joint relate to monolithic lengths of only 122 m.

General arrangement and articulation (Fig. 5.16(a))
Anchored at crown of arch and both abutments

Reference

BAXTER, J.W, GEE, A.F. and JAMES, H.B. (1965) Gladesville Bridge. *Proceedings of the Institution of Civil Engineers*, **30**, 489–530. (Paper No. 6860.)

Fig. 5.16   Gladesville Bridge, Sydney, Australia, (a) articulation diagram; (b) plan; (c) general view.

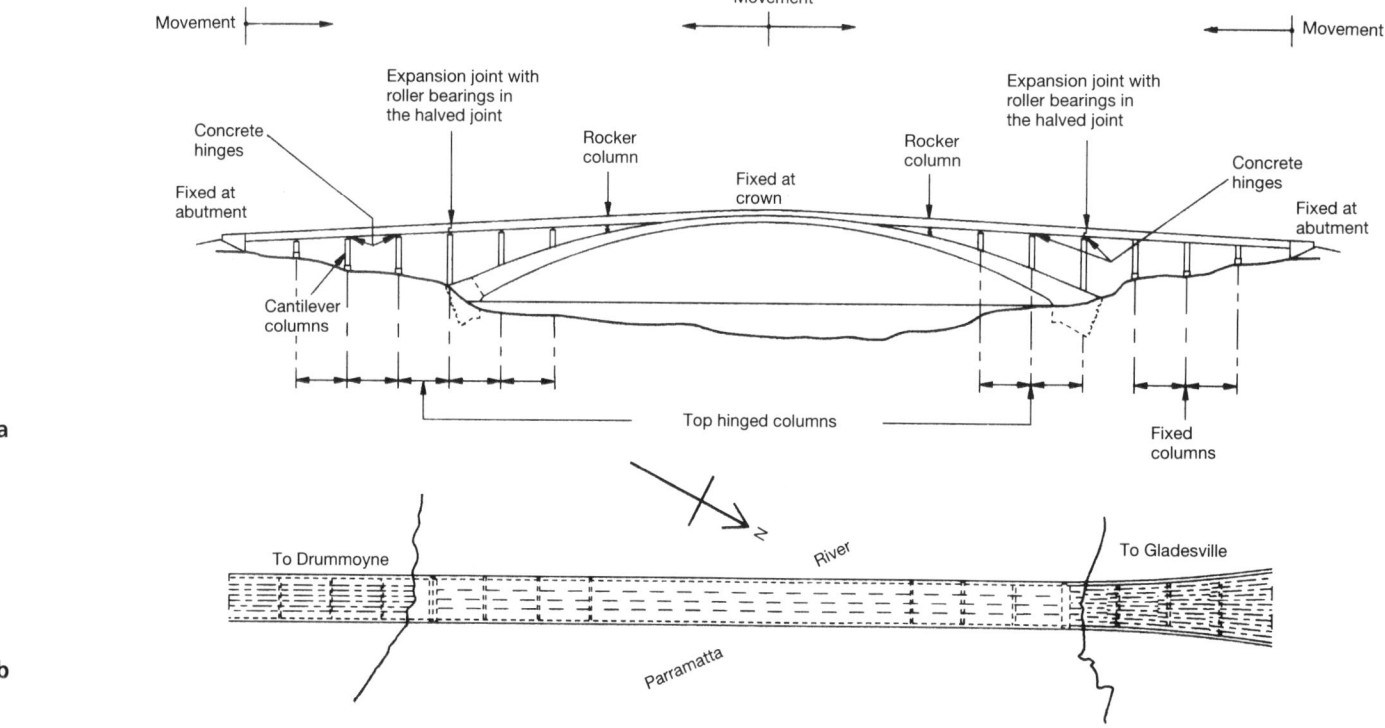

5.17 Tasman Bridge, Hobart, Tasmania

Length   West approaches totalling 110 m approx.; 22 main spans totalling 1025 m; 12 east approach spans totalling 256 m

Horizontal alignment   Straight on main spans

Width   17.4 m overall

Superstructure type   Main spans: precast prestressed concrete I-beams with *in-situ* infilled concrete deck

Substructure type   Main spans: rigid hollow concrete column bents beneath the three navigation spans and flexible concrete column bents beneath the side spans, all comprising precast concrete segments stressed together by Macalloy bars

Foundation type   Bored cylinder piles 1.37 m diameter

Bearings   Side spans: rocker bearings

Ends of anchor spans: Stainless steel roller bearings beneath side and anchor spans

Navigation span piers: rocker bearings

Centre drop in span: stainless steel roller bearings at one end and rocker bearings at the other

Movement joints   Comb type joints at ends of anchor spans and at one end of the centre drop in span

Note   The combined flexibility of the column bents and foundation piles is assumed

General arrangement and and articulation (Fig. 5.17; also Figs 2.9 and 2.10.)   Anchored at east and west abutments at navigation span piers

Reference

NEW, D.H., LOWE, J.R. and READ J. (1967) The superstructure of the Tasman Bridge, Hobart. *The Structural Engineer*, **45**(2), 81–90.

Fig. 5.17   Tasman Bridge, Hobart, Tasmania, articulation diagram.

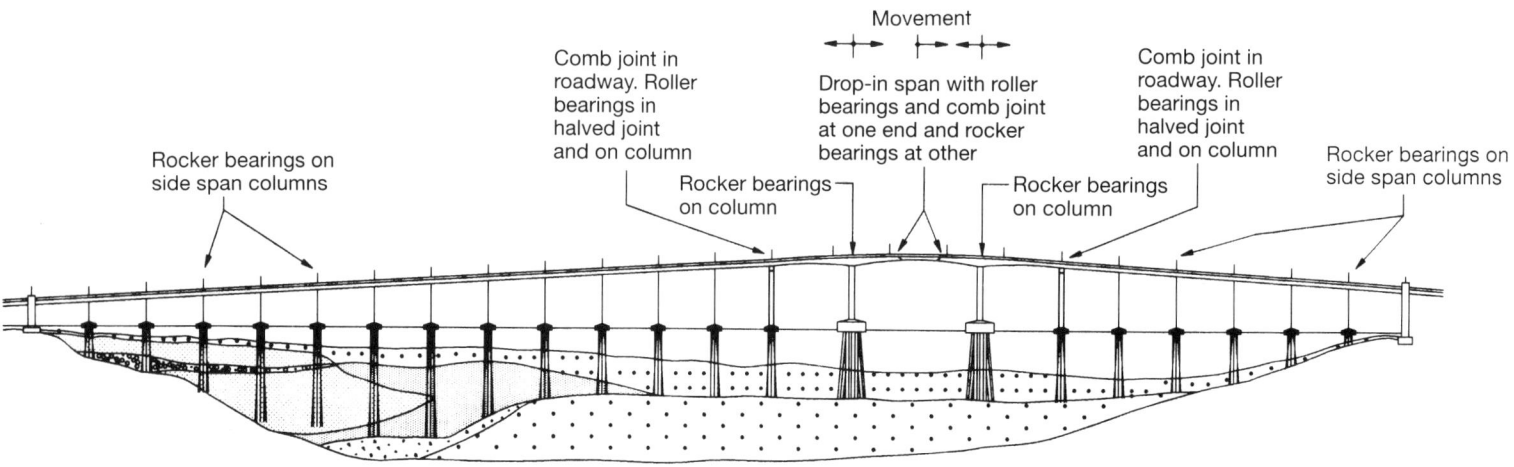

Movement

Comb joint in roadway. Roller bearings in halved joint and on column

Drop-in span with roller bearings and comb joint at one end and rocker bearings at other

Comb joint in roadway. Roller bearings in halved joint and on column

Rocker bearings on side span columns

Rocker bearings on column

Rocker bearings on column

Rocker bearings on side span columns

West

East

5.18   Bowen Bridge, near Hobart, Tasmania

Length   10 spans totalling 976 m
Horizontal alignment   Straight
Width   21.4 m overall
Superstructure type   Twin continuous precast prestressed concrete box girders connected by an *in-situ* concrete strip
Substructure type   Twin hollow reinforced concrete piers integral with the foundation caissons
Foundation type   River spans: 14 m overall diameter hollow concrete caissons on rock
Bearings   Rocker bearings at the central pier and pairs of roller bearings under each box elsewhere with 600 mm movement capacity. Vertical load capacity of each bearing 1600 t
Movement joints   Concertina type (Repco–Honel joints) with 450 mm movement capacity at the abutments
Note   This bridge with 15.2 m vertical clearance was constructed 5 km upstream from Tasman Bridge (vertical clearance 45.7 m)

General arrangement and articulation (Fig. 5.18(a))
Anchored at the central river pier

References

1. LESLIE, J.A. and ALLEN, J.B. (1983) *Design of the Bowen Bridge, Hobart.* Engineering Conference Newcastle: Engineering towards the 21st century; Newcastle on the Hunter, IEAUST 18–23 April, Barton, pp. 163–70.
2. ANON. (1984) Bowen Bridge. *Constructional Review*, **57**, (1).
3. ANON. (1986) The Bowen Bridge with precast prestressed segments at Hobart, Australia. *L'Industria Italiana del Cemento*, No. 11.
4. IABSE (1983) Bowen Bridge, Hobart (Australia). IABSE Structures C-28/83, periodical **4**, 90–91.

Fig. 5.18   Bowen Bridge, near Hobart, Tasmania, (a) articulation diagram; (b) plan; (c) general view.

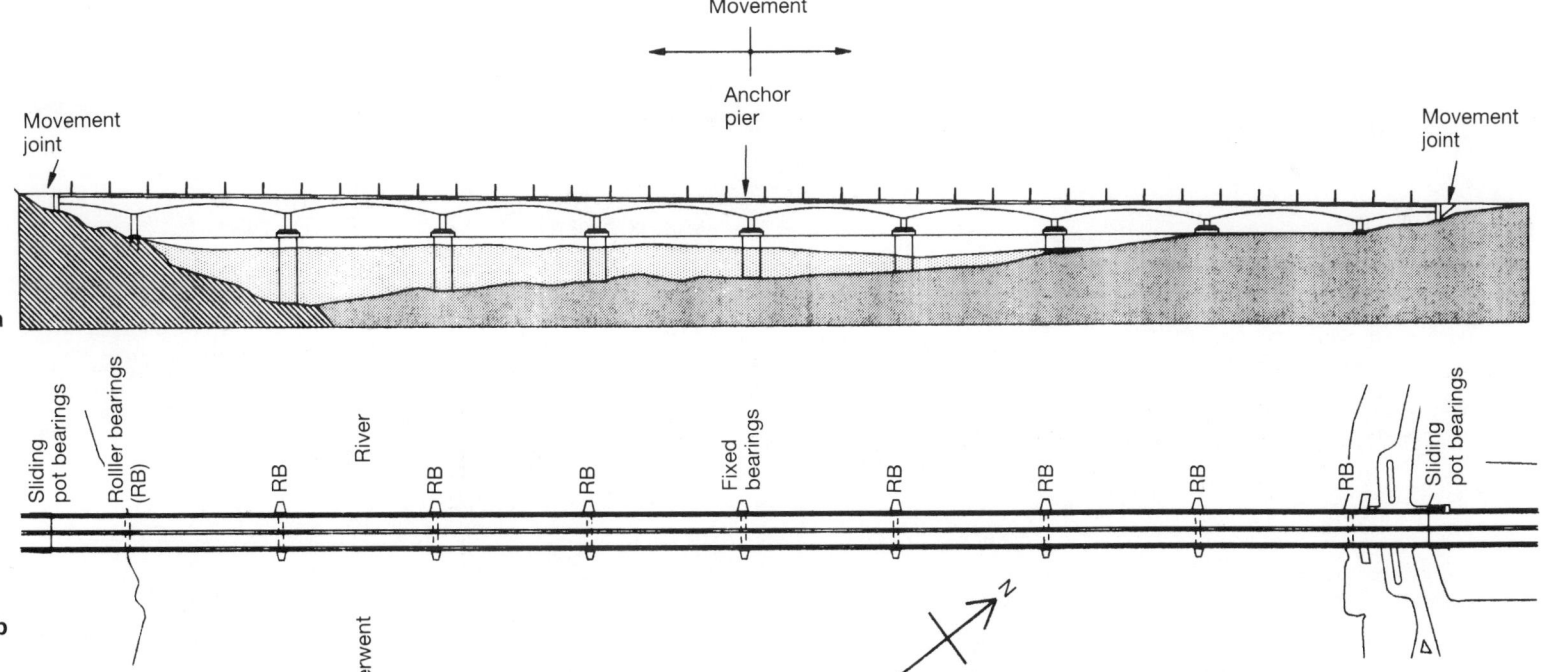

5.19 Ahmad Shah Bridge, Temerloh, Malaysia

Length   Eight approach spans totalling 129 m; two main spans totalling 273 m; seven approach spans totalling 106 m

Horizontal alignment   Straight

Width   14.5 m overall (designed for future widening to 23.5 m overall)

Superstructure type   Main spans: twin continuous steel box girders composite with reinforced concrete deck
   Approach spans: simply supported precast prestressed concrete inverted T-beams with *in-situ* reinforced concrete slab deck

Substructure type   Main spans: reinforced concrete box abutments and solid reinforced concrete wall river pier
   Approach spans: reinforced concrete raking leg frame piers

Foundation type   Main spans: bored cylinders 1 m diameter
   Approach spans: Franki piles 0.52 m diameter

Bearings   Main spans: steel rocker bearings at east abutment and the river pier; Hi-load roller bearings at west abutment
   Approach spans: elastomeric bearings

Movement joints   Main spans (at abutments): comb type joints in carriageways and stainless steel sliding plate joints in footways. Approach spans: buried joints at cross-wall supports.

General arrangement and articulation (Fig. 5.19)   Main spans anchored at east abutment with flexible river pier (spring stiffness 0.6 t/mm)

Reference

Lee, D.J. and Wallace, A. (1977) Ahmad Shah Bridge, Malaysia. *Proceedings of the Institution of Civil engineers*, **62**, 89–118. (Paper No. 7980.)

Fig. 5.19   Ahmad Shah Bridge, Termerloh, Malaysia, (a) articulation diagram; (b) plan; (c) main pier rocker bearing; (d) viaduct support bent; (e) general view.

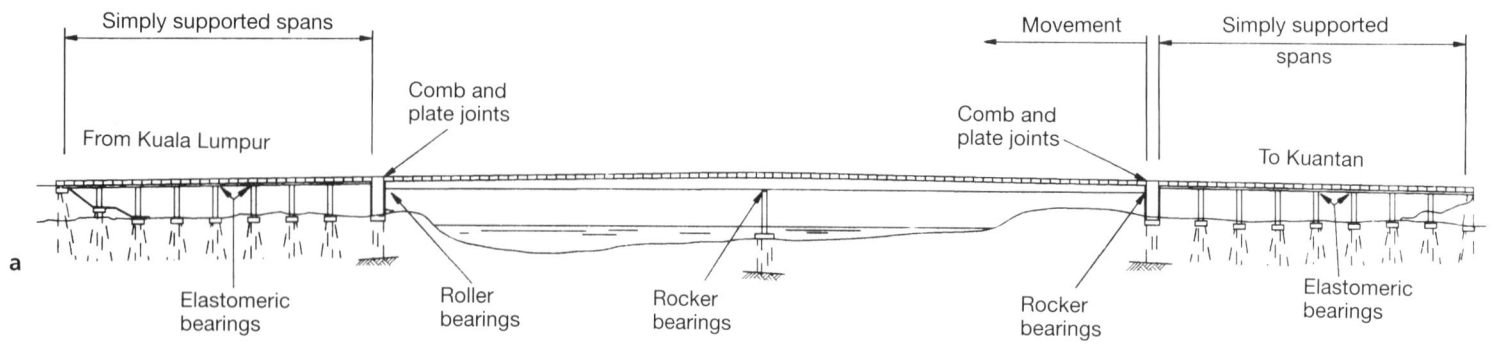

a

Simply supported spans

From Kuala Lumpur

Comb and
plate joints

Movement

Simply supported
spans

Comb and
plate joints

To Kuantan

Elastomeric
bearings

Roller
bearings

Rocker
bearings

Rocker
bearings

Elastomeric
bearings

b

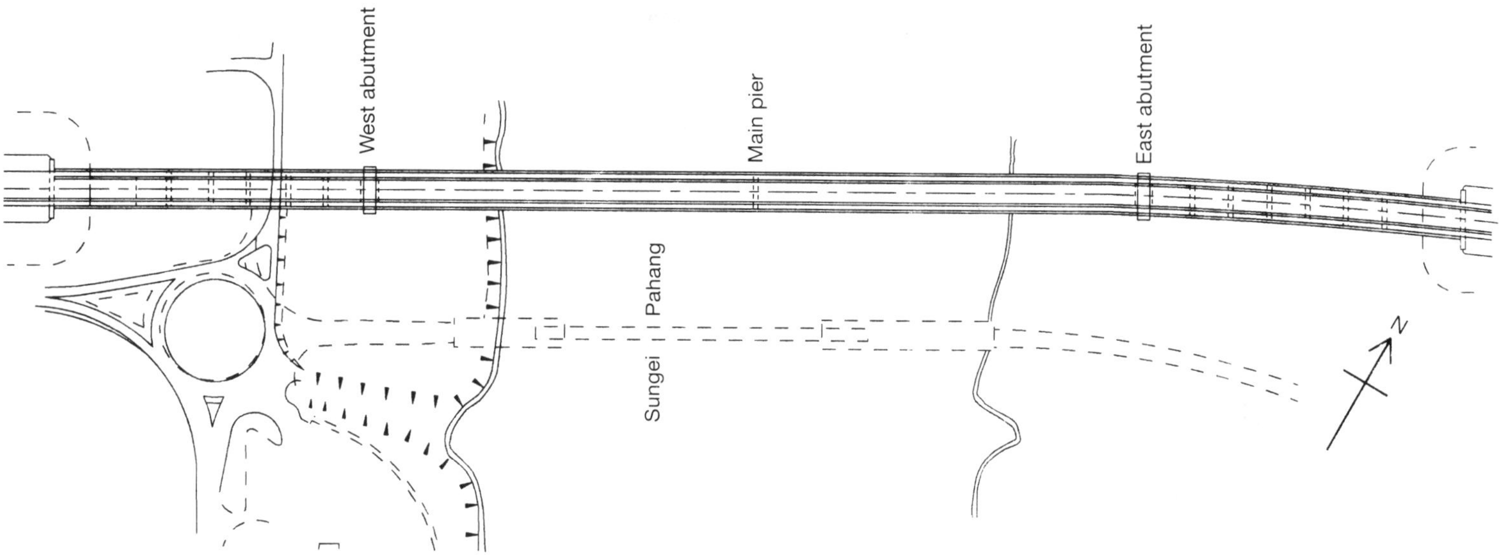

West abutment

Main pier

Sungei   Pahang

East abutment

N

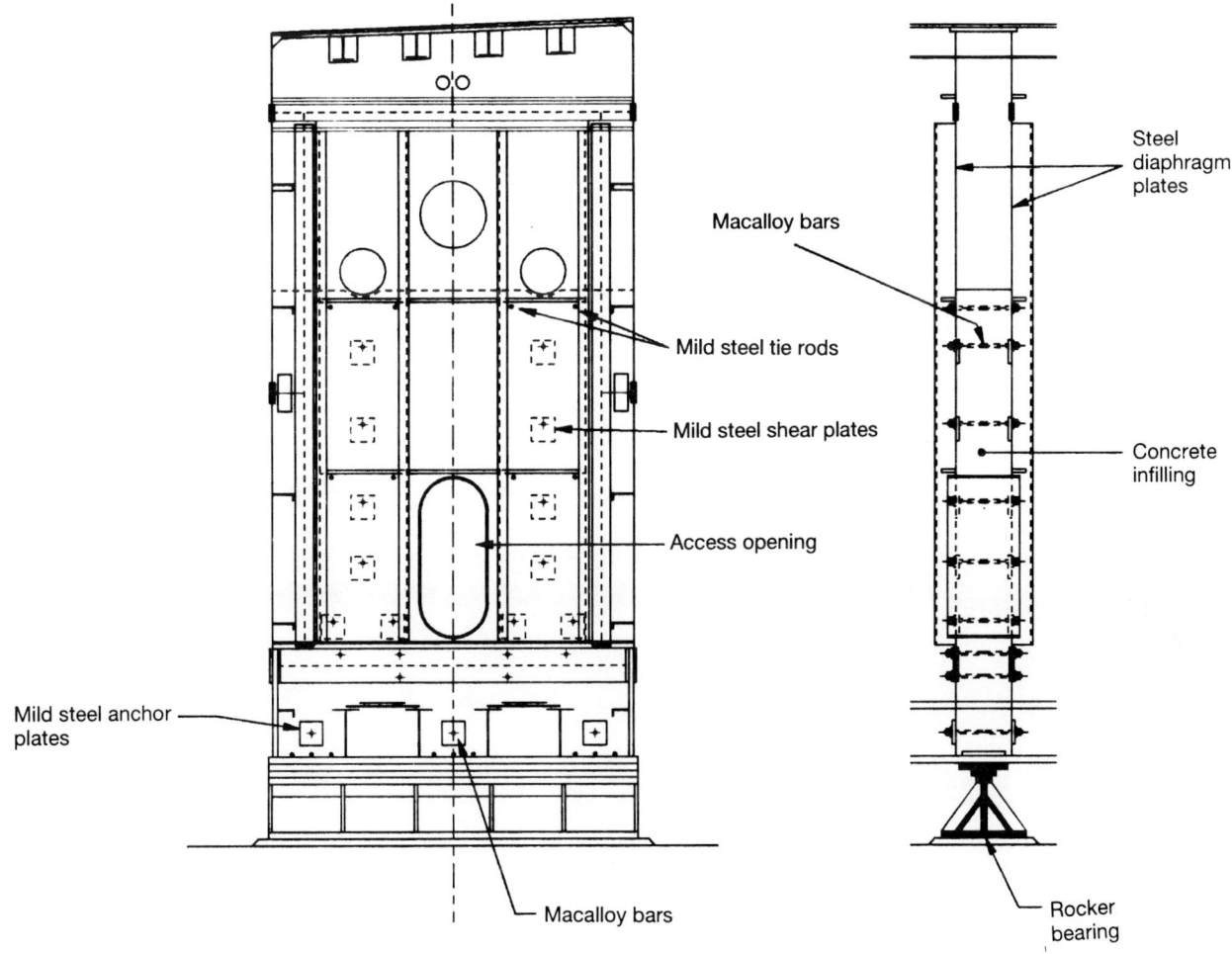

Steel diaphragm plates

Macalloy bars

Mild steel tie rods

Mild steel shear plates

Concrete infilling

Access opening

Mild steel anchor plates

Macalloy bars

Rocker bearing

c

Elevation

Section

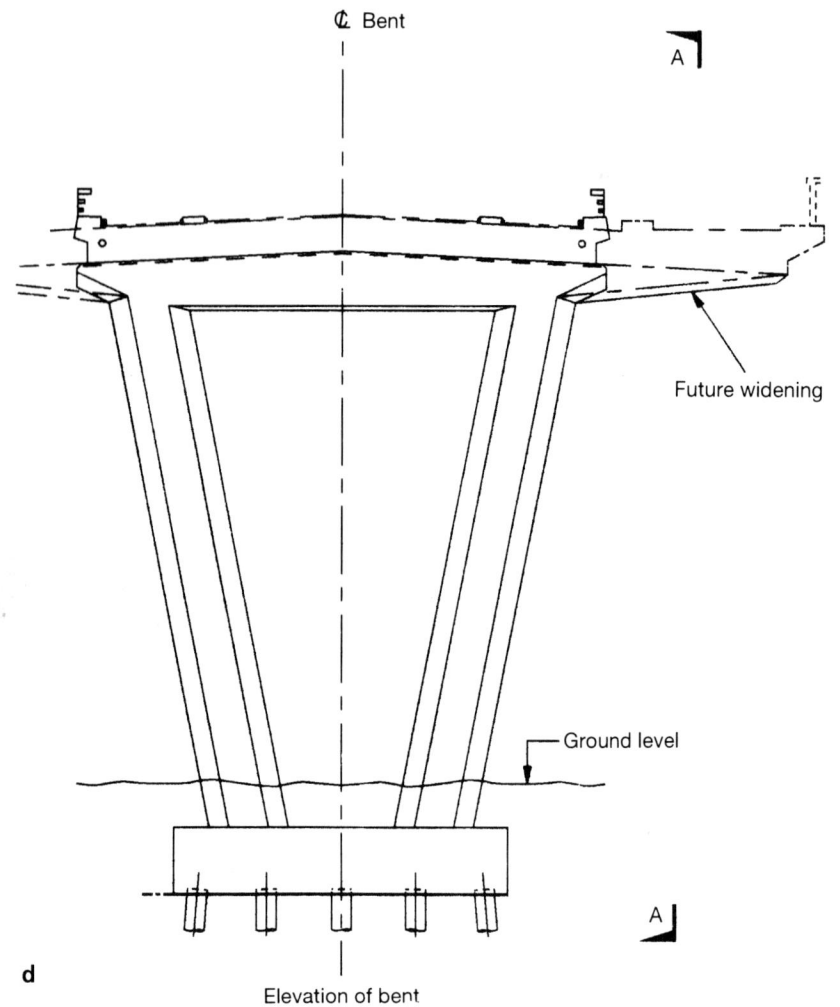

C Bent

A

Future widening

Ground level

d

Elevation of bent

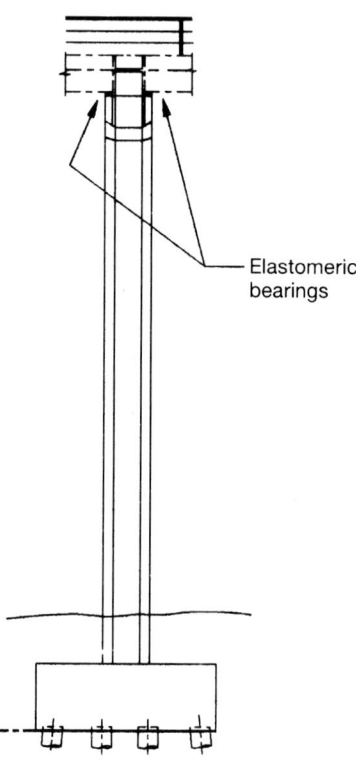

Elastomeric
bearings

Elevation A–A

5.20   Adhamiyah (14th Ramadhan) Bridge, Baghdad, Iraq

Length   Four main spans totalling 370 m

Horizontal alignment   Straight

Width   30.22 m overall

Superstructure type   Continuous steel spine box with cantilevers and composite with the concrete deck. The main span is supported by a single plane asymmetrical cable system

Substructure type   Reinforced concrete piers cantilevering up from the pile caps

Foundation type   Bored cylinders 0.75 m, 1.2 m and 2.0 m diameter

Bearings   Link bearings at the cable anchorage piers
Independent rocker bearings under both the deck and the pylon at the main river pier.
   Pot bearings at the other pier and the south abutment

Movement joints   Cushion joints at both ends of the structure

General arrangement and articulation (Fig. 5.20)   Anchored at the main river pier

Reference

LEE, D. J. and WALLACE, A. (1984) Adhamiyah Bridge, Baghdad. *The Structural Engineer*, **62A** (1), 11–25.

Fig. 5.20   Adhamiyah (14th Ramadhan) Bridge, Baghdad, Iraq, (a) articulation diagram; (b) plan; (c) prestressed bearing link; (d) general view.

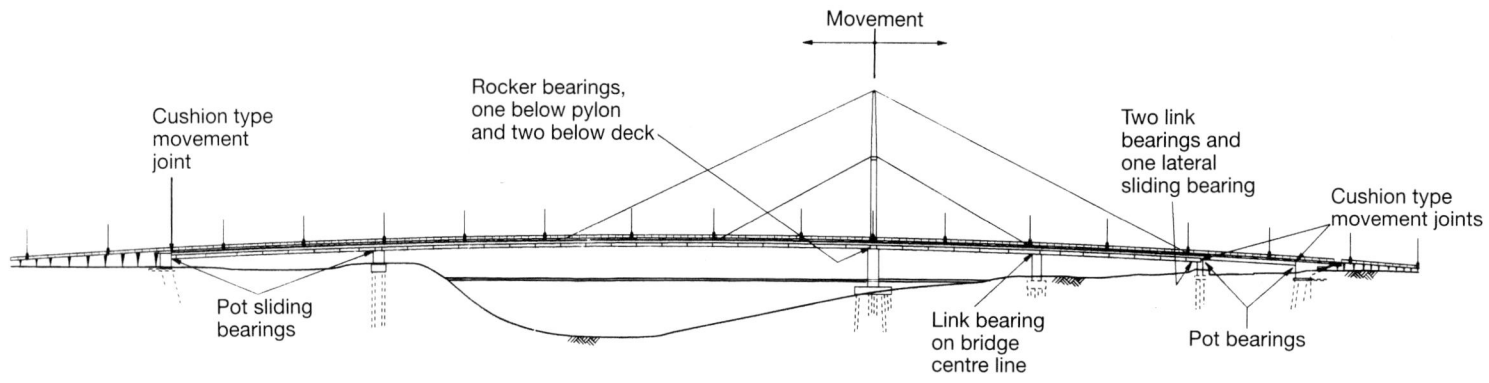

a

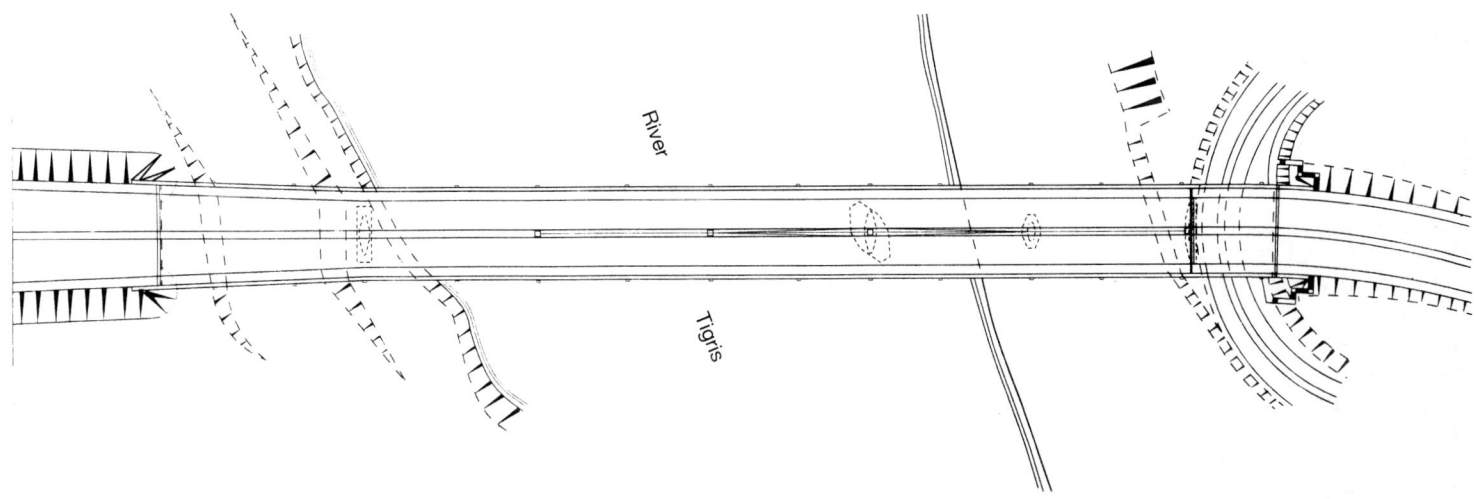

b

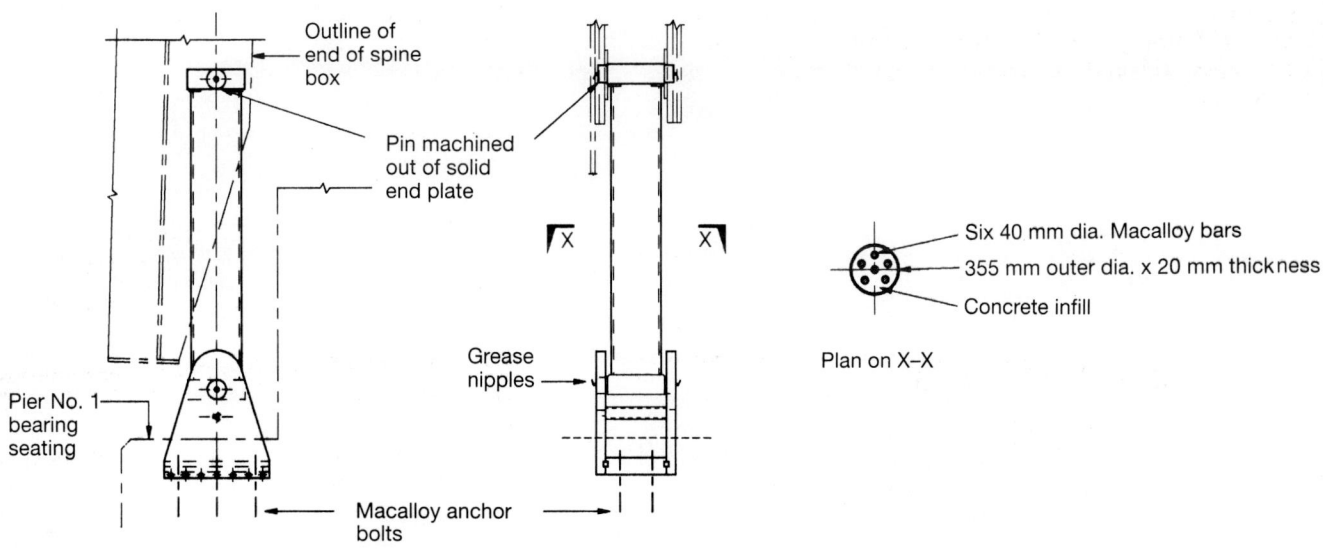

Outline of
end of spine
box

Pin machined
out of solid
end plate

Pier No. 1
bearing
seating

Macalloy anchor
bolts

Grease
nipples

Six 40 mm dia. Macalloy bars

355 mm outer dia. x 20 mm thickness

Concrete infill

Plan on X–X

c

PART TWO

Particular Features of Selected Bridges

5.21   Severn Bridge, UK

Three span steel suspension bridge totalling 1597 m.
Anchored at the abutments with longitudinal movement joints at both towers.
The vertical loads at the ends of the suspended structure are carried by hinged tubular links at the abutments and by hinged I-section links at the towers.
At the abutments, lateral steel bearings located on the bridge centre line incorporate long high tensile steel screwed rods to provide longitudinal restraint while permitting end rotations of the deck by flexure. At the towers the centrally located lateral bearings slide against bearing brackets projecting from the tower portal beam below the bridge deck.
Demag rolling leaf type roadway joints are used in the roadway and sliding plate joints in the footway and cycle tracks.

References

1. ROBERTS, Sir G. (1968) Severn Bridge: design and contract arrangements. *Proceedings of the Institution of Civil Engineers*, **41**, 1–48. (Paper No. 7138.)
2. GOWRING, G.I.B. and HARDIE, A. (1968) Severn Bridge: foundations and substructure. *Proceedings of the Institution of Civil Engineers*, **41**, 49–67. (Paper No. 7119.)
3. HYATT, K.E. (1968) Severn Bridge: fabrication and erection. *Proceedings of the Institution of Civil Engineers*, **41**, 69–104. (Paper No. 7084.)

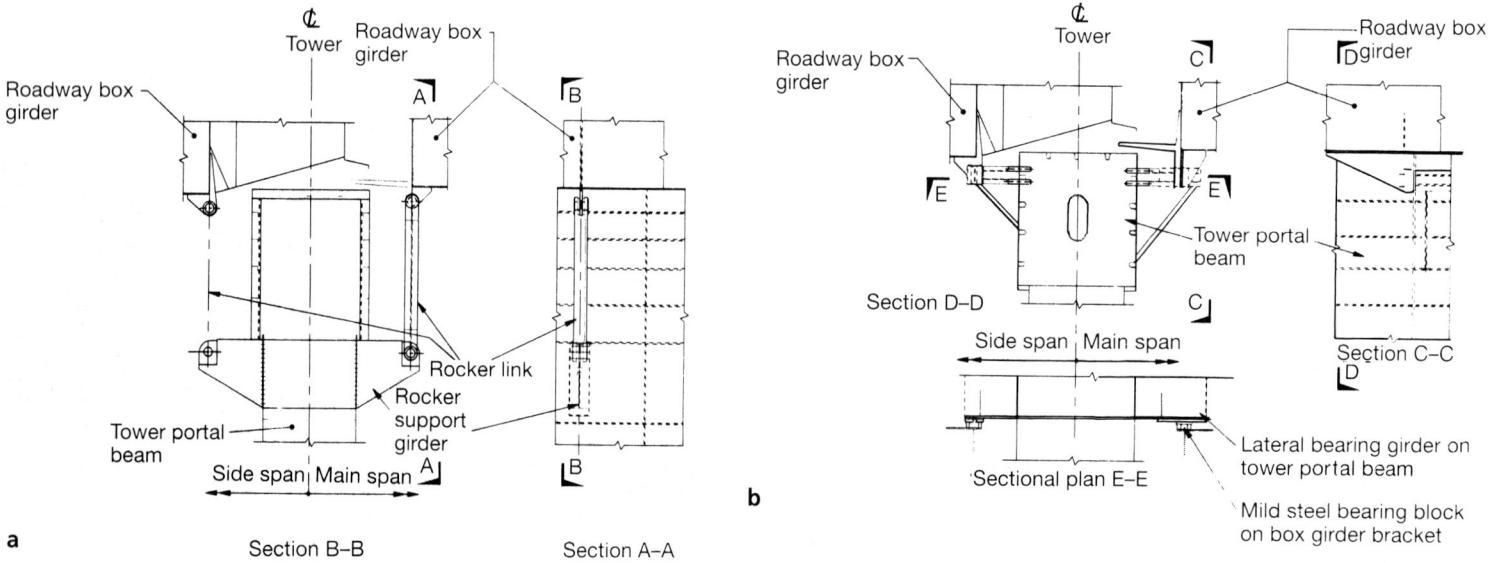

Fig. 5.21 Severn Bridge, UK, (a) rocker link bearing at tower; (b) lateral bearing at tower; (c) general view.

5.22 Farö Bridge (main spans), Denmark

Three-span symmetrical double plane cable-stayed steel bridge with orthotropic plate deck.

The connection between the deck and the tower comprises:

1. vertical pendulum links, allowing differential movements of the deck and the tower;
2. horizontal plate springs controlling longitudinal movements of the deck;
3. vertical sliding bearings on the tower allowing the deck to deflect.

Hydraulic cylinders incorporated in the pendulum links are cross-linked by fluid lines between opposite sides of the deck so that they are effectively rigid when the deck is subjected to pure torsion. Pressure transducers and a datalogging system allowing simultaneous recordings of torsional movements, wind velocity and direction, traffic loading etc. are expected to provide valuable data.

The approach spans are supported on long movement Glacier spherical bearings.

Reference

OSTENFIELD, K.H. and HASS, G. (1984) Torsional fixation of girders in cable suspended bridges, *International Association of Bridge and Structural Engineering*, 12th Congress, Vancouver, Final Report, pp. 787–92.

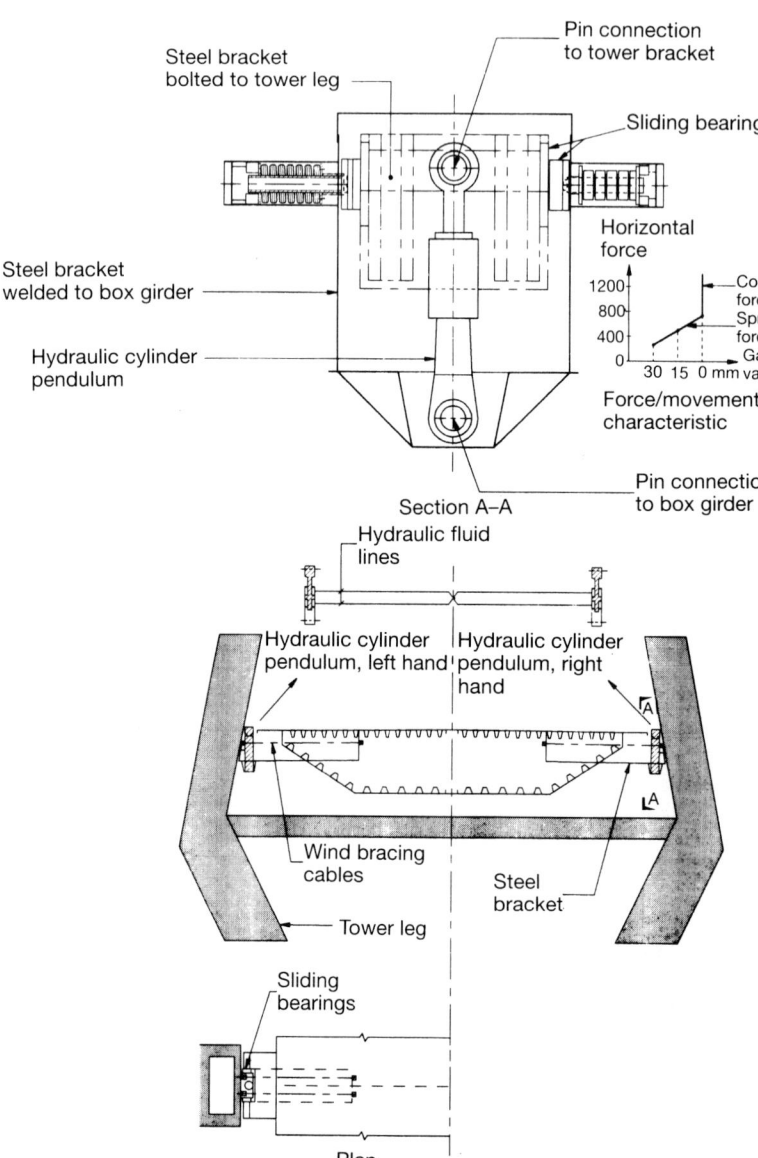

Fig. 5.22 Farö Bridge, Denmark.

5.23   Penang Bridge (main spans), Malaysia

Three-span symmetrical double plane cable-stayed concrete bridge.
Pot sliding bearings permitting free longitudinal and transverse movement
and a lateral wind lock at the end piers only.
Steel lateral bearings only between girder and inside faces of tower legs.
Modular expansion joints at each end pier.

Reference

CHIN FUNG KEE and McCABE, R. (1990) Penang Bridge project:
superstructure design and construction. *Proceedings of the Institution of Civil
Engineers*, **88**, 551–70.

Fig. 5.23   Penang Bridge (main spans), Malaysia.

5.24   Tsing Yi North Bridge (main spans), Hong Kong

Three-span twin box prestressed concrete bridge.

The superstructure is anchored longitudinally and transversely at the west main pier and is supported on pot bearings on top. Pot sliding bearings are used at the east main span and sidespan piers. Seismic forces are resisted by an upstand key on the pier caps with vertical metal bearings reacting against the bottom corners of the box girders. Compressible disc spring stacks are incorporated in these bearings to accommodate shortening of transverse beams which connect the boxes over the upstand key.

Reference

DENTON-COX, R.A. and WEIR, K.L. (1989) Tsing Yi North Bridge: planning and design. *Proceedings of the Institution of Civil Engineers*, **86**, 471–89. (Paper No. 9364.))

Fig. 5.24   Tsing Yi North Bridge (main spans), Hong Kong

5.25   Benjamin Sheares Bridge, Singapore

Twenty-three-span main viaduct totalling 1750 m between abutments
consisting of reinforced concrete trestle frames supporting precast
prestressed concrete box table top and cantilever decks interconnected by
simply supported precast prestressed I-beams with *in-situ* concrete decks.
The multicurved structure varies in width from 27 m to 39 m.

References

1. ANON. (1979) Package Deal Viaduct pays off in Singapore. *New Civil
    Engineer*, 19 July, 36–8.
2. ANON. (1983) East Coast Parkway, Singapore. *IABSE Periodica* 4/1983.
    Structures C.28/83)

Fig. 5.25   Benjamin Sheares Bridge, East Coast Parkway, Singapore,
(a) articulation diagram; (b) general view.

a

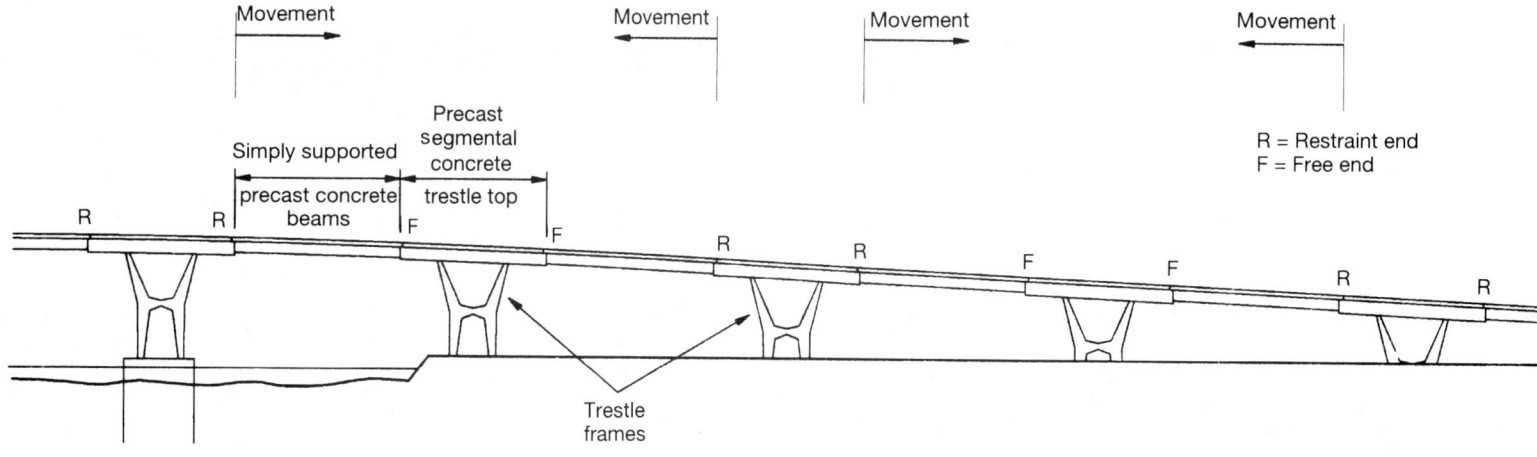

5.26 Singapore Mass Rapid Transport System

Viaducts formed from precast prestressed concrete box girders, typically 25 m simply supported spans carrying ballasted track with continuous welded rail.

Several types and concepts of bearings were adopted.

phase 1:   four laminated rubber bearings per beam with horizontal loads resisted by shear pins, one end fixed and one end guided;

phase 1A: four pot bearings per beam, fixed at one end and partially guided at the other;

phases IIA and IIB: three pot bearings, one fixed at one end and two partially guided at the other.

Phase IA bearings were easier to set than those in Phase I and will be easier to remove and replace. The Phase II arrangement allows for rotation of the crossheads due to differential settlement of piles.

For the annual 12°C temperature range in Singapore expansion devices were not considered necessary over the free bearings.

Reference

COPSEY, J.P., HULME, T.W., KRAFT, B. and SRIPATHY, P. (1989) Singapore Mass Transit System: design. *Proceedings of the Institution of Civil Engineers*, **86**, 667–770. (Paper No. 4429.)

Fig. 5.26   Singapore Mass Rapid Transport System.

5.27 Huntley's Point Overpass, Sydney, NSW, Australia

Length Seven unequal spans totalling 178 m
Horizontal alignment Asymmetric horizontal curve 107 m minimum radius
Width 13.11 m overall
Superstructure type Continuous cast *in-situ* prestressed concrete spine box with cantilevered slab outer decks
Substructure type Reinforced concrete piers and box abutments
Foundation type Spread footings on rock. Anchor span abutment rock anchored through precast concrete double hinges.

Bearings Steel rocker bearings at anchor span pier 1750 t capacity on the inside and 750 t on the outside of the curve, elastomeric bearings elsewhere.
Movement joints Steel comb type joints at both abutments
Note The superstructure was jacked clear of the elastomeric bearings at the end of each construction stage so that they were not required to take the full creep and shrinkage movements
General arrangement and articulation (Fig. 5.27) Anchored at anchor span pier

Reference
LEE, D.J. (1965) Huntley's Point Overpass. *Journal of the Reinforced Concrete Association – Structural Concrete*, **2**(12), 521–31.

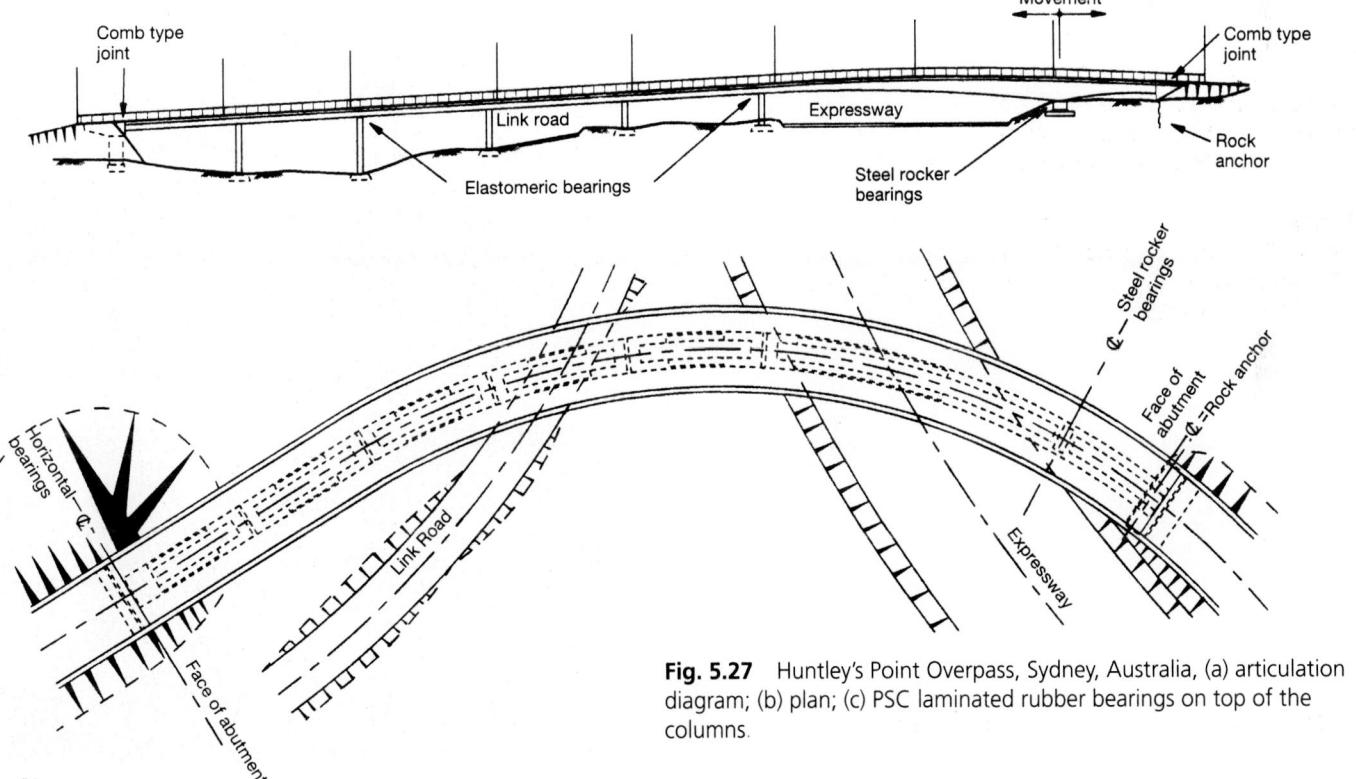

Fig. 5.27 Huntley's Point Overpass, Sydney, Australia, (a) articulation diagram; (b) plan; (c) PSC laminated rubber bearings on top of the columns.

5.28   Tarban Creek Bridge, Sydney, NSW, Australia

Length   214 m
Horizontal alignment   Straight
Width   29.2 m overall
Superstructure type   116 m span concrete arched portal frame of precast U-shaped units with *in-situ* deck.

Fig. 5.27c

5.29 Gateway Bridge (main spans), Brisbane, Australia

Three-span precast prestressed concrete spine box girder with cantilevers. The main river piers have twin flexible columns integral with the superstructure and base to accommodate movements of the 260 m centre span. Mageba joints are located in the side spans near the side piers.

Fig. 5.28 Tarban Creek Bridge, Australia, general view.

5.30   Jindo and Dolsan Bridges, Korea

Three-span symmetrical double plane cable-stayed steel box girders with orthotropic plate decks.
Tie-down plate link bearings and roller plate expansion joints are used at the abutments.

Reference

TAPPIN, R.G.R. and CLARK, P.J. (1985) Jindo and Dolsan Bridges: design. *Proceedings of the Institution of Civil Engineers*, **78**, 1281–300. (Paper No. 8994.)

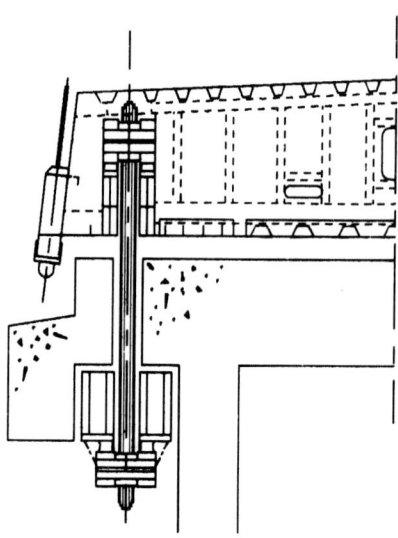

Fig. 5.27   Jindo and Dolsan Bridges, Korea, backstay rocker link detail.

Appendix: Summary of bridge examples

		Superstructure (2.3.1)				Super/substructure (2.3.2)		Substructure (piers) (2.3.3)					Substructure/foundation (2.3.4)		Joint types (3.2 to 3.4)					Bearing types (4.3.1 to 4.3.11)								
	Fig. no.	Slightly supported	Continuous single structure	Continuous twin structures with gap	Suspended span	Articulated connection	Rigid connection	Stiff	V-frame	Flexible wall	Flexible twin wall	Pinned top and bottom	Articulated connection	Rigid connection	Buried	Comb and sliding plate	Reinforced elastomeric	Rolling leaf	Railway	Roller	Rocker	Knuckle pin	Spherical	Link	Sliding	Pot	Elastomeric	Concrete hinge
Hammersmith Flyover, London, UK	5.1		×				×	×					×		×	×				×								
Medway Bridge, Rochester, Kent, UK	5.2		×	×		×		×						×		×				×								×
Mancunian Way, Manchester, UK	5.3		×			×	×	×					×			×									×	×		
Westway, London, UK	5.4		×			×	×	×					×	×		×		×				×			×	×		
London Bridge, UK	5.5	×			×	×		×						×		×								×				×
Tyne and Wear Metro Bridge, N106, UK	5.6		×			×		×						×					×	×					×			
Orwell Bridge, Ipswich, UK	5.7		×			×		×						×			×			×	×				×			
Redheugh Bridge, Newcastle, UK	5.8		×			×				×				×		×									×	×		×
Foyle Bridge, Northern Ireland	5.9			×		×		×		×				×		×				×		×			×	×		

Text reference: 2.3.1   2.3.2   2.3.3   2.3.4   3.2 to 3.4   4.3.1 to 4.3.11

	Text reference	2.3.1	2.3.2	2.3.3			2.3.4		3.2 to 3.4		4.3.1 to 4.3.11							
	Fig. no.	Superstructure	Super/substructure	Substructure (piers)			Substructure/foundation		Joint types		Bearing types							
Kylesku Bridge, Highland Region	5.10	×	×		×			×	×		×							
Torridge Bridge, Bideford, UK	5.11	×	×	×				×	×		×							
River Dee Viaduct, Clywd, Wales	5.12	×	×	×		×		×	×					×	×			
Narrows Bridge, Perth, Australia	5.13	×	×				×	×		×		×	×					
Westgate Bridge Approaches, Melbourne	5.14	×	×				×	×		×			×					
Commonwealth Avenue Bridge, Canberra	5.15	×	×	×				×	×			×						
Gladesville Bridge, Sydney, Australia	5.16	×	×	×		×	×	×	×			×						×
Tasman Bridge, Hobart, Tasmania	5.17	×	×			×		×	×			×	×					
Bowen Bridge, Tasmania	5.18	×	×	×				×				×	×					
Ahmed Shah Bridge, Malaysia	5.19	×	×		×	×	×	×				×	×				×	
Adhamiyah Bridge, Baghdad, Iraq	5.20	×	×	×				×		×			×	×	×	×		
Severn Bridge, UK	5.21									×				×				
Farö Bridge, Denmark	5.22													×	×	×		
Penang Bridge, Malaysia	5.23													×	×			
Tsing Yi North Bridge, Hong Kong	5.24														×	×		
Benjamin Sheares Bridge, Singapore	5.25																×	
Singapore MRT System	5.26																	
Huntley's Point Overpass, Sydney	5.27	×	×	×				×	×				×				×	
Tarban Creek Bridge, Sydney, Australia	5.28																	
Gateway Bridge, Brisbane, Australia	5.29					×												
Jindo and Dolsan Bridges, Korea	5.30														×			

Index